About IFPRI

The International Food Policy Research Institute (IFPRI), established in 1975, provides research-based policy solutions to sustainably reduce poverty and end hunger and malnutrition. The Institute conducts research, communicates results, optimizes partnerships, and builds capacity to ensure sustainable food production, promote healthy food systems, improve markets and trade, transform agriculture, build resilience, and strengthen institutions and governance. Gender is considered in all of the Institute's work. IFPRI collaborates with partners around the world, including development implementers, public institutions, the private sector, and farmers' organizations.

About IFPRI's Peer Review Process

IFPRI books are policy-relevant publications based on original and innovative research conducted at IFPRI. All manuscripts submitted for publication as IFPRI books undergo an extensive review procedure that is managed by IFPRI's Publications Review Committee (PRC). Upon submission to the PRC, the manuscript is reviewed by a PRC member. Once the manuscript is considered ready for external review, the PRC submits it to at least two external reviewers who are chosen for their familiarity with the subject matter and the country setting. Upon receipt of these blind external peer reviews, the PRC provides the author with an editorial decision and, when necessary, instructions for revision based on the external reviews. The PRC reassesses the revised manuscript and makes a recommendation regarding publication to the director general of IFPRI. With the director general's approval, the manuscript enters the editorial and production phase to become an IFPRI book.

Pulses for Nutrition in India

Changing Patterns from Farm to Fork

Edited by Devesh Roy, P. K. Joshi, and Raj Chandra

A Peer-Reviewed Publication

International Food Policy Research Institute
Washington, DC

International Food Policy Research Institute
1201 Eye Street, NW, Washington, DC 20005, USA
Telephone: 202-862-5600

DOI: https://doi.org/10.2499/9780896292567

Library of Congress Cataloging-in-Publication Data

Names: Roy, Devesh, editor. | Joshi, P. K., editor. | Chandra, Raj, editor.
Title: Pulses for nutrition in India : changing patterns from farm to fork / edited by Devesh Roy, P. K. Joshi, and Raj Chandra.
Description: Washington, DC : International Food Policy Research Institute, 2017. | Includes bibliographical references and index.
Identifiers: LCCN 2017015672 | ISBN 9780896292567
Subjects: LCSH: Legumes as food—India. | Legume industry—India. | Legumes—India. | Nutrition policy—India.
Classification: LCC TX558.L4 P85 2017 | DDC 641.3/5650954—dc23 LC record available at https://lccn.loc.gov/2017015672

Cover Design: Anne C. Kerns, Anne Likes Red
Project Manager: Patricia Fowlkes, IFPRI
Book Layout: BookMatters, Berkeley, CA

Contents

Tables

Figures

Box

Abbreviations and Acronyms

AAY	Antyodaya Anna Yojana (grain)
AEZ	agri-export zones
AIC	Akaike Information Criteria
AICPIP	All India Coordinated Pulses Improvement Project
AICRP	All India Coordinated Research Project
APMC	Agricultural Produce Marketing Committee
ASF	animal source foods
ASI	Annual Survey of Industries
ASTI	Agricultural Science and Technology Indicators
AVRDC	World Vegetable Center
CACP	Commission for Agricultural Costs and Prices
CAGR	compound annual growth rates
CES	Consumer Expenditure Survey
CFTRI	Central Food Technological Research Institute
CI	convergent innovation
CIAE	Central Institute of Agricultural Engineering
CMIE	Center for Monitoring of Indian Economy
CSO	Central Statistical Organization
CSP	community service providers
DES	Directorate of Economics and Statistics

DSIR	Department of Scientific and Industrial Research
EOU	export-oriented unit
EPZ	Export Processing Zone
FCDS	Food Characteristics Demand System
FDI	foreign direct investment
FHP	farm harvest price
FICCI	Federation of Indian Chambers of Commerce and Industry
FLD	front line demonstrations
FPO	farmer producer organizations
FPVS	farmers' participatory varietal selection
GBPUAT	Gobind Ballav Pant University of Agriculture and Technology
GDP	gross domestic product
GST	Generalized System of Taxes
HS	Harmonized System
IARI	Indian Agricultural Rcsearch Institute
ICAR	Indian Council of Agricultural Research
ICARDA	International Center for Agricultural Research in the Dry Area
ICRISAT	International Crop Research Institutes for the Semi-Arid Tropics
IIPR	Indian Institute of Pulses Research
IPGA	India Pulse Growers Association
IPM	integrated pest management
IRF	impulse response functions
ISOPOM	Integrated Scheme on Oilseeds, Pulses, Oil Palm, and Maize
KVK	Krishi Vigyan Kendra
LMIC	low- and middle-income countries
LOU	livestock output unit
MAGIC	Multi-Parent Advanced Generation Inter-Cross
MCCHE	McGill Centre for the Convergence of Health and Economics
MFE	meat-fish-eggs

MOPU	More Pulse
MPCE	monthly per capita consumption expenditure
MSP	minimum support price
MYMV	mung bean yellow mosaic virus
NABARD	National Bank for Agriculture and Rural Development
NAFED	National Agricultural Cooperative Marketing Federation of India
NARS	national agricultural research systems
NFSM	National Food Security Mission
NPDP	National Pulses Development Programme
NSSO	National Sample Survey Office
OGL	Open General License
OH	outside home
ORMAS	Odisha Rural Development and Marketing Society
PDKV	Panjabrao Deshmukh Krishi Vidyapeeth
PIP	Pulse Innovation Partnership
PM	powdery mildew
PMKSY	Pradhan Mantri Krishi Sinchayee Yojana (Prime Minister Agriculture Irrigation Plan)
PPP	public-private partnerships
PSB	phosphate solubilizing bacteria
RDI	recommended dietary intake
RKVY	Rashtriya Krishi Vikas Yojana
SAUS	state agricultural universities
SEZS	Special Economic Zone Schemes
SFAC	Small Farmers Agribusiness Consortium
SMD	sterility mosaic virus
SRR	seed replacement rate
TCS	Tata Consultancy Services
TE	triennium ending
TMO	Technology Mission on Oilseeds

TNAU	Tamil Nadu Agricultural University
VAM	vesicular arbuscular mycorrhiza
VAT	value-added tax
VECM	vector error correction model
WCO	World Customs Organization
WPI	Wholesale Price Index

Foreword

The persistent scarcity in pulses and some other commodities amid lagging nutrition, particularly in times of climate change, has reopened the debate whether more balanced, inclusive, and climate-smart food and agricultural policies would ensure food and nutrition security for millions of the poor in India. Historically, faced with dire situations regarding the availability of food and a high dependence on food aid and imports, policy makers rightfully have focused on ensuring basic availability of food grains to India's population. The policy stance toward agriculture and food markets has been cereal-centric for a long time, with the unintended consequence of neglecting many competing crops such as oilseeds and pulses. The degree of cereal-based food security achieved painstakingly over the decades faced new challenges in making agriculture more nutrition-sensitive and climate-smart.

In this context, pulses in India occupy a unique place: this "meat" for the poor provides several environmental benefits as well. India is home to millions of poor and malnourished people, therefore the country's pulses sector urgently needs to rebound. The government has adopted a mission mode with actions on several fronts. The agonizing trends of lagging nutrition indicators and stagnant growth in agriculture call for significant rebalancing of the portfolio where pulses play an important role. With the dust settling on food price spikes in India to some extent, the time for a serious and informed response to tackling food security and nutrition on a sustained basis is critical. This book focuses on this extremely important topic and considers issues in India's pulses sector from different perspectives, highlighting the need for a holistic approach to resurrect the sector. The implications for food security, nutrition, and sustainable development are clear for a crop as important as pulses.

Without taking any idealist stand on policies to foster growth in the pulses sector, this book assesses India's policies and institutions and offers a possible way forward given the government's steadfast focus on leading the pulses sector to regain a place of prominence. The book offers an ideal platform for analysis where all aspects of the pulses value chain have been analyzed. Not surprisingly, the analyses show a dependence of outcomes on the interactions of different components.

Taking a historical view, this edited volume points to the direness of India's initial situation. If with policy reforms and the right mix of policies (on technology, spatial integration, and markets), India could come out of the problems in the case of cereals, choosing the right policies for pulses (particularly in the face of food price spikes, nutrition, and environmental challenges) is a compelling one. This book is an important and timely contribution in this direction as it highlights the high level of interdependence among different components of the pulses value chain with efficacy of policies depending on the response of each. The analysis clearly states that there are significant gains to be made from learning from varied experiences across regions and crops, including the adequate role of the private sector.

Arvind Subramanian
Chief Economic Adviser
Department of Economic Affairs
India, Ministry of Finance

Foreword

In recent years, policy makers and researchers all over the world have been turning more attention to nutrition. In India, this is leading to a shift in policy focus away from basic staples such as wheat, rice, and corn to more protein-rich crops such as pulses, which also contribute to environmental sustainability and help to mitigate the effects of climate change through balancing nitrogen levels in soil. *Pulses for Nutrition in India: Changing Patterns from Farm to Fork* is in part the product of discussions, debates, and research on this topic among the authors, policy makers, and other stakeholders.

In India, rising food prices are forcing policy makers to rethink the country's cereal-centric agricultural policies and how they affect food security and nutrition. This book presents robust analyses based on research and evidence drawn from years of food policies that privilege cereals over other crops and how such policies have affected consumption, trade, and processing as well as other indicators. The findings suggest that India's longtime policies aimed at grain self-sufficiency have stifled the pulses sector—historically so important in India—and likely harmed food security and nutrition outcomes. These policies were designed and first put into place shortly after independence, when the country was strikingly food insecure.

India must accelerate its progress in reducing hunger, malnutrition, and food insecurity. It is time for the food-policy framework to be revisited. This book builds the case for policy reform aimed at resurrecting India's pulses sector. The book also explores the probable consequences of not realigning agricultural policies to level the playing field for key nongrain crops such as pulses. Equipped with this information, India's policy makers will be better able to design effective food policies and to shape a new agricultural paradigm based on nutrition and health rather than on calories.

Shenggen Fan
Director General, IFPRI

Acknowledgments

This book is a product of significant effort involving several colleagues from National Agricultural Research System (NARS) and international organizations, along with various policy makers. We greatly acknowledge the support of donors that include the Indian Council of Agricultural Research (ICAR) and the Swiss Agency for Development and Cooperation. This work was also undertaken as part of the CGIAR Research Program on Agriculture for Nutrition and Health (A4NH). The opinions expressed here belong to the authors, and do not necessarily reflect those of A4NH or CGIAR. We'd also like to thank the government departments and ministries, and the scientists from the National Agricultural Research System, especially the Indian Institute of Pulses Research who provided the intellectual inputs that shaped the study and added value in many ways.

The editors and authors thank Gershon Feder and Corinne Garber, chair and secretary respectively, of the Publications Review Committee at the International Food Policy Research Institute (IFPRI), who directed the book through the review process. The editors would also like to thank Tracy Brown (senior editor); Patricia Fowlkes, the IFPRI project manager, for the manuscript; and Anne Kerns, who designed the book's cover.

Celebrating the International Year of Pulses 2016, the International Food Policy Research Institute (IFPRI) held several roundtables, workshops, and conferences that helped immensely in developing *Pulses in India: Changing Patterns from Farm to Fork*. The covering of different themes on pulses (viz., production, consumption and environmental services, prices and markets, technologies, policies and institutions, and human health) provided us the spectrum of expert knowledge and debate needed for a book of this type.

Chapter 1

INTRODUCTION

Devesh Roy, P. K. Joshi, and Raj Chandra

Food security has long been the mainstay of public policy in India. Only in the 1970s did the country begin to come out of the shadow of severe food scarcity with its history of acute famines. Even in the mid-1960s, its import dependency concerning cereals was as high as 16 percent, which coupled with the country's small export base also had serious balance-of-payment repercussions. Consequently, India's agricultural policy has always aimed to maximize agricultural production and achieve self-sufficiency, particularly in cereals.

Several technological, infrastructural, and institutional changes were implemented to increase agricultural production. Even with those advances, however, rapid growth in the country's population meant its land and other resources remained under great pressure to meet food needs. Therefore, high-yielding crops such as rice and wheat, which could lead to increased food production, received greater attention. The late 1960s and 1970s witnessed a spectacular growth in the production of these crops with the introduction of Green Revolution technologies.[1] The spread of these technologies did not confer equal benefits across crops or regions of the country. Vast areas in the eastern and western regions remained untouched by the technologies. An additional challenge was that with the adoption of technology and new access to irrigation, wheat and rice recorded significant growth at the cost of displacing other crops, particularly coarse cereals, pulses, and oilseeds.

1 The Green Revolution in India refers to a period when agriculture principally of cereals increased its yields due to improved agronomic technology in the early 1960s. This led to a significant increase in food production of rice and wheat, especially in such areas as Punjab, Haryana, and western Uttar Pradesh. It was brought about by the spread of higher-yielding varieties of the crops supported by the increased use of chemical fertilizers and irrigation.

The Case for Pulses

Pulses, long considered "the poor man's meat" because of their protein profile, occupy a unique place in India. India ranks as the country with the world's largest number of malnourished people. Because of this, coupled with the country's high incidence of vegetarianism, the future of pulses is of special significance to India's large poor population. Pulses are also among the crops that have been adversely affected by the dominance of cereals over the past several decades, so it is clear that agricultural policy is implicated in both the nutritional challenges and, as we hope to show, in potential solutions.

Naturally, the cost of food is significant to India's poorest population, and in recent years the country has suffered a persistent problem of food price inflation. Many basic foods exhibit a higher average rate of inflation than the overall Wholesale Price Index (WPI); Sonna et al. (2014) observe that cereals, pulses, milk, fruits and vegetables, meat-fish-eggs (MFE), and sugar all exhibited higher average rates of inflation. At the same time, the relative price increases have been driven largely by protein-rich foods (Gokarn 2011), whose cost has been rising uniformly at a faster rate than the cost of other foods. The most common contributors to this inflation are milk and fish (Mishra and Roy 2011); since 2005, the average inflation rates for pulses and MFE have been higher than the rates for the composite food WPI. Taken together, pulses, MFE, milk, and milk products constituted around 30 percent of the total food expenditure according to the 66th Round (2009–2010) of the national survey by the National Sample Survey Office (NSSO), but they were responsible for approximately 42 percent of food inflation since 2005 (Sonna et al. 2014; Mishra and Roy 2011). It is important to note that although the cost of pulses has been an important driver of food price increases, pulses continue to be cheaper than several of the other sources of protein, including animal source food (ASF).[2]

Pulses, in fact, constitute the most common source of noncereal protein in India, where the frequency of pulse consumption is higher than that of any other protein source. Among Indian consumers, pulses contributed nearly 10 percent of the protein consumed in 2011–2012. This is closely linked with the Indian tradition of vegetarianism—widespread reliance on plant-based diets—a common and deeply ingrained dietary pattern that dates back at least 2,500 years. Indians constitute about 70 percent of the world's population of vegetarians. About three-quarters of Indian vegetarians are lacto-vegetarians—that is, although they do not consume meat or eggs, they

2 ASF comprises meat, fish, eggs, and dairy.

have no prohibitions for milk or other dairy products. And up to 25 percent of India's vegetarians are lacto-ovo vegetarians—they too do not eat meat, but they consume eggs as well as dairy products (Rammohan, Awofeso, and Robitaille 2012). As a source of protein, pulses of different types are eaten across all of India's regions. Other principal sources of protein include MFE. A more recent figure on vegetarianism is available from the baseline data for the Census of India. Based on this data, there still is significant presence of vegetarians in the population. In 2014, of those 60 years and older, nearly 33 percent report as vegetarians in both urban and rural areas. Also, the incidence of vegetarianism is slightly higher among women than men in all age groups in urban and rural households. In the age group that is above 30, 29.8 percent of women report as vegetarians compared with 28.2 percent among men (Census of India 2014).

Pulses are consumed equally by India's rich and poor as it is one of the less expensive sources of protein (Mohanty and Satyasai 2015). Around 89 percent of consumers eat pulses at least once a week, while the corresponding number for eating fish or chicken/meat once a week is only 35.4 percent (IIPS and ORC Macro 2007). Pulses complement the staple cereals in people's diets with proteins, essential amino acids, vitamins, and minerals. They contain 22 percent to 24 percent protein, almost twice the amount of protein found in wheat and three times that found in rice. Pulses have a unique nutritional profile consistent with several dietary composition factors thought to assist with weight control. They also contain several antinutrients that play a role in energy regulation. Pulses are high in fiber, relatively low in energy density (1.3 kcal per gram), and a good source of digestible protein (average of 7.7 grams of protein per half cup). Pulse carbohydrates are slowly digested (McCrory et al. 2010). The amount of protein in pulses is 17 percent to 35 percent on a dry weight basis (Boye, Zare, and Pletch 2010). Jukanti et al. (2012) provide the nutrition benefits of chickpea specifically and present it as a good source of carbohydrates and protein, with protein quality in particular better than other pulses. Chickpea has significant amounts of all the essential amino acids except the sulfur-containing types, which can be complemented by adding cereals to daily diet as cereals are rich in sulfur-containing amino acids. Jukanti et al. (2012) list chickpea as an affordable source of protein, carbohydrates, minerals and vitamins, dietary fiber, folate, beta-carotene, and health-promoting fatty acids.

Pulses also possess advantages for soil health and farming sustainability. Because of India's diverse agroclimatic conditions, pulses are grown in various parts of the country throughout the year. Their growth reduces soil pathogens and fixes nitrogen in the soil, which enhances soil productivity, improving the yields of crops that follow their harvest. In this way, pulses play a vital role in crop rotation and intercropping. Studies show that because of these factors, the yield of crops that follow pulses can increase by 20 percent to 40 percent (Pande and Joshi 1995). The changes in soil fertility brought about by pulse cultivation have been assessed for different crops, including maize (Dwivedi et al. 2015; Peoples et al. 2009). Because pulse farming itself requires less use of fertilizer, pesticide, and irrigation than many other crops, pulses are also an environmentally friendly crop group (Reddy, Bantilan, and Mohan 2013). India currently suffers from excessive chemical usage, and the government is saddled with a huge burden in fertilizer subsidies (equal to 1.5 percent of GDP on average). Moreover, heavy pesticide use presents food safety issues. Finding relief from these stresses is of first-order importance in India.

Portfolio of Pulses Grown in India

Several types of pulses are grown in India, as illustrated in Box 1.1. Despite being the leading producer of pulses, India has been consistently unable to meet its own domestic demand for the food. For several reasons, pulse production has been increasingly disadvantaged over the years and has become relatively less profitable compared with cereal production in areas of reasonable fertility and access to irrigation:

1. The Green Revolution pushed pulse cultivation away from irrigated areas to rainfed areas, where nearly 87 percent are now grown. This reliance on rain, however, makes pulses a risky crop.
2. Technology development has been far more extensive and more yield-improving for cereals than for pulses.
3. In addition, being protein-rich also makes pulses more prone to different types of pests and diseases.

On the policy side, the system of a minimum support price (MSP) with procurement may make growing and selling pulses comparatively less risky for farmers. Where MSP is announced at the time of sowing and procurement occurs, the government agrees to buy all the grain (rice and wheat) that is offered for sale at that price, removing all the price risk. MSP currently applies to paddy rice, wheat, five coarse grains, four pulses, eight oilseeds, cotton, jute, tobacco, and sugarcane.[3] The announced MSP for pulses (with insignificant procurement) has risen consistently over time. Notwithstanding these increases, the MSP in pulses still serves only as a benchmark price and remains far below the market price; for pigeon pea, for example, in 2015 the MSP notional price was *less than one-fourth* of the market price. For this reason, pulse farmers must continue relying on traders for their sales rather than selling to government procurement. NSS data, however, show that in the case of wheat and rice, only 6 percent of farmers gain access to government procurement, casting doubts on how many pulse farmers would gain (and to what extent) from a larger pulse procurement program if one were enacted. The effectiveness of such a system remains in question and the issue certainly warrants further research.

Pulse milling is almost a widespread industry in the Indian subcontinent, but it has not received the scientific and technological support from the government necessary to modernize it, unlike other food-processing industries, such as rice and wheat milling (Banerjee and Palke 2010).

Challenge: The Decline of Pulses over Time

Because of the relatively disadvantaged position of pulses in comparison with cereals, over the past 56 years, pulse production has risen by only 32 percent as compared with a roughly 280 percent increase in cereal production over the same period. The crop yield from pulses has shown a similar trend, gaining only by 25 percent as compared with a 211 percent gain in cereals (Srivastava, Sivaramane, and Mathur 2010). Moreover, pulse yields have been widely variant across the different areas where the crop is grown.

At the same time, the land area devoted to pulses marginally decreased, from 24 million hectares during the triennium ending in 1975 to 23 million hectares during the triennium ending in 2005. This shrinkage was due to

3 The MSP for pulses has been increased by more than 50 percent over the past five years and has often been boosted with bonuses. For example, the agriculture ministry announced up to a 6 percent increase in MSP in 2014–2015, including a bonus of 200 rupees per quintal. With the increase, the MSP of black matpe reached 4,625 rupees per quintal for the crop year 2015–2016 (July–June) as against 4,350 rupees per quintal the previous year. Over the past four years, the increase in MSP was a massive 87 percent for tur, 71 percent for black matpe, and 63 percent for green gram. Among rabi pulses for MY 2014–2015, the MSP for chickpea was fixed at 3,100 rupees per quintal and the MSP for lentil at 2,950 rupees per quintal; these prices represented a modest increase over the 2013–2014 levels of 3,000 rupees and 2,900 rupees per quintal, respectively, but they represented a massive increase of 76 percent and 58 percent, respectively, since 2010–2011 (NCAER 2014).

BOX 1.1 **Supply and demand characteristics of different types of pulses in India**

India is the world's largest producer and consumer of pulses. Major pulses grown in India include chickpea or Bengal gram, pigeon pea or red gram, lentil, black matpe, mung bean or green gram, lablab bean, moth bean, horse gram, pea, grass pea or khesari (*Lathyrus sativus*), cowpea (*Vigna unguiculata*), and broad bean or faba bean (*Vicia faba*). Popular pulses in India are chickpea, pigeon pea, green gram, black matpe, and lentil. Pulses are mostly grown in two seasons: (1) the warmer, rainy season or *kharif* (June–October), and (2) the cool, dry season or *rabi* (October–April) (Gowda et al. 2013). Chickpea, lentil, and dry peas are grown in the *rabi* season, while pigeon pea, black matpe, green gram, and cowpea are grown during *kharif*. Among the various pulses, chickpea dominates, claiming a more than 40 percent share in production of all pulses grown, followed by pigeon pea (18–20 percent), green gram (11 percent), black matpe (10–12 percent), lentil (8–9 percent), and other legumes (20 percent) (IIPR 2011).

The major pulses—chickpea, pigeon pea, lentil, green gram, and black matpe—account for nearly 80 percent of total pulse production in India. India's total production, in turn, accounts for 33 percent of world production by area and 22 percent of world production by volume. By area, India's production makes up 90 percent of global production of pigeon pea, 65 percent of chickpea, and 37 percent of lentil; this corresponds to 93 percent, 68 percent, and 32 percent of the global production of these pulses, respectively, by volume (FAO 2011). Among all pulses, lentil is the most actively traded (about 25 percent of world production of lentil is internationally traded) (Reddy and Reddy 2010). Lentil is an important *rabi* pulse crop, next only to gram, and it is distinctive in being the only pulse grown in India with a net exportable surplus (all other pulse trade has a significant net import reliance). Pulses with their local names are presented Table B1.1.

On the consumption side, the annual per capita consumption of pulses declined between 1993–1994 and 2004–2005 (from 9.44 kilograms to 8.82 kilograms) and then rose again by 2011–2012 (to 9.6 kilograms), a consumption pattern that has been mirrored by each of the major pulse crops: pigeon pea, gram (split), green gram, and black matpe. As a share of total food expenditure, pulses represent about 5 percent. Among pulses, pigeon pea is the most heavily consumed, making up more than 30 percent of total pulse expenditure, although the type of pulse most demanded varies significantly across states. The major chickpea-consuming states are Haryana, Punjab, and Rajasthan. The major pigeon pea–consuming states are Andhra Pradesh, Karnataka, and Maharashtra, while the major green gram–consuming state is Gujarat. In Assam, Bihar, and West Bengal, lentil is the pulse in greatest demand. Black gram is prominent in Tamil Nadu and Uttar Pradesh (Reddy 2004).

In terms of prices, the wholesale price index has been higher for pulses compared with cereals and oilseeds in the past two decades. The variation in price has also been relatively higher for pulses, with black matpe, green gram, and pigeon pea experiencing the greatest increase in price over time.

TABLE B1.1 English, local, and scientific names of the pulses

English name	Local name	Scientific name
Pigeon pea, red gram	Arhar, tur	*Cajanus cajan*
Chickpea, bengal gram, garbanzo bean	Chana	*Cicer arietinum*
Lentil	Masoor	*Lens culinaris*
Green gram, mung bean	Moong	*Vigna radiate*
Black matpe, black gram, black matpe bean	Urad, udid, urad bean	*Vigna mungo*
Pea	Matar	*Pisum sativum* var. *arvense*
Grass pea	Khesari	*Lathyrus sativus*
Yellow pea	[No local name]	*Lathyrus aphaca*
Lablab bean	[No local name]	*Lablab purpureus*
Moth bean	Moth	*Vigna aconitifolia*
Horse gram, madras gram	Kulti	*Dolichos uniflorus*
Broad bean, faba bean		*Vicia faba*
Cowpea		*Vigna unguiculata*

Source: Authors.

farmers' shifting over to nonpulse crops, for which the government has made irrigation and infrastructural facilities available (Gowda et al. 2013). Only recently, possibly because of price increase, the area allocation to pulses recovered to 26.3 million hectares in 2011–2012 (Gowda et al. 2013) and receded again to 25 million hectares in 2013–2014 (Mohanty and Satyasai 2015). The decline of the pulses sector in India is reflected in three broad facts:

1. **Per capita consumption.** Consumption of pulses has fallen over time and currently stands at levels below those attained in the 1980s. Consumption fell continuously from the 1980s through the 2000s, although it has been improving again over the past few years

(2012–2014), a period when production significantly increased to 17–18 million metric tons.[4]

2. **Inflation.** There has been a persistent increase in pulse prices, resulting in accessibility issues for the poor.
3. **Imports.** Imports of pulses have sharply increased and have been expanding on the extensive margin.

Although pulses may become marginalized, one of the book's themes is that this set of crops offers a wealth of opportunities, and its potential in India has not yet been fully exploited. Advances in food and crop technology have not been fully deployed to advantage in the case of pulses, opportunities for inducing more efficient value chains have not been taken up, and the potential efficacy of price management has not been adequately studied regarding the different pulses crops.

Better price management could involve combining support prices with procurement (although the likely effectiveness of minimum support prices policies combined with procurement for pulses entails potential challenges and requires further study) and seeking to ensure better transmission of prices to the farmgate through direct purchases along with processing. Such a combined approach could help arrest the consumption slide that these crops have experienced since the 1990s. In a comprehensive study, Tiwari, Gowen, and Mckenna (2011) have shown that pulses are nutritionally diverse crops that can be successfully used as a food ingredient or a base for innovative product development. Today, new options have become available in food processing, including technologies for processing whole pulses, techniques for fractionating pulses into ingredients that preserve their functional and nutritional properties, and other potential applications to incorporate pulses into new food products.

The remainder of this chapter presents some facts that form the background motivation for this book: the production and consumption patterns of India's pulse sector, the global context of trade in pulses and the position of India therein, the nutritional and environmental characteristics that make pulses a salient crop in meeting human needs, and the different initiatives for the pulses sector taken up by the government of India.

4 Throughout this chapter, "tons" are "metric tons."

Pulse Production and Consumption

The production of pulses in India tends to fluctuate quite significantly. Table 1.1 presents recent production statistics for different pulses, which show that there is little consistency. For example, consider chickpea production: while it was 8,833 thousand tons in 2012–2013, it was merely 7,170 thousand tons just two years later. Its share in total pulse production moved from 48 percent to just 42 percent.

Since production has been volatile and restricted within a narrow range, a demand-supply gap has been a constant feature of pulses since early 2000, resulting in stubbornly high prices.[5] Figure 1.1 illustrates this price phenomenon using unit values from the NSSO Consumer Expenditure Survey (CES). There is a distinct rising trend in pulses prices and a falling trend in consumption over two decades. The demand elasticity has been estimated to be quite high for pulses, particularly for the poor households, by different studies (−0.7 or higher).

TABLE 1.1 Production of pulses in India, 2012–2013 to 2014–2015

	2012–2013		2013–2014		2014–2015	
Pulses per year	Production of pulses (thousands of metric tons)	Share in total pulse production (%)	Production of pulses (thousands of metric tons)	Share in total pulse production (%)	Production of pulses (thousands of metric tons)	Share in total pulse production (%)
Pigeon pea	3,023	16	3,170	16	2,780	16
Chickpea	8,833	48	9,530	48	7,170	42
Green gram	1,186	6	1,610	8	1,510	9
Black matpe	1,947	11	1,700	9	1,870	11
Lentil	1,134	6	n.a.	n.a.	n.a.	n.a.
Other pulses	2,220	12	3,780	19	3,870	23
Total pulses	18,343	100	19,780	n.a	17,200	n.a.

Source: India, Ministry of Agriculture, New Delhi various years, Directorate of Economics and Statistics (DES).
Note: n.a. = not applicable.

5 References to *demand-supply gap* here and elsewhere in the book pertain to the gap between domestic demand and domestic production.

FIGURE 1.1 Pulse prices and consumption (all-India level, 1988–2011)

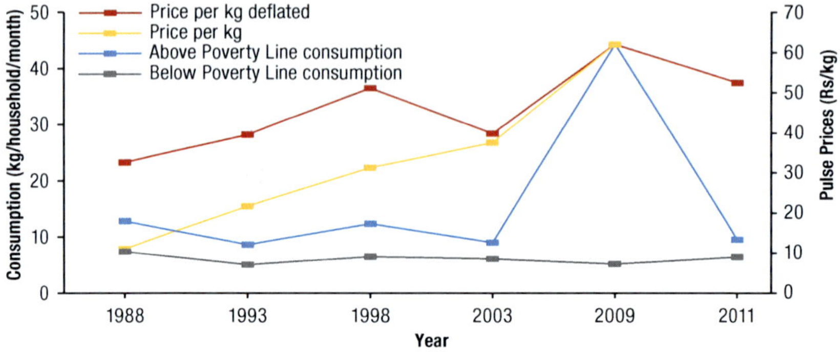

Source: National Sample Survey Office consumption expenditure data.

Production in India: Regional Trends

Although pulse production in India has remained stagnant since the late 1980s (with the exception of the last three years), significant changes have occurred at the subnational level. Crowding out by expanded cereal production, discussed earlier, has led to pulse production centers moving from the eastern to the western region and from the northern to the southern region. Apart from pushing pulses to more marginal environments, this has also meant that for many pulses, areas of production ceased to be same as the areas where they are most consumed. This has made the role of the supply chain especially important.

Some crowding out of pulses over time has also occurred due to the cultivation of noncereal crops. Rapid expansion of soybean, for example, has had adverse effects on the areas planted in certain *kharif* pulses, including pigeon pea and green gram. Even though pulse yields have not shown much dynamism, important technological developments have improved yield, mostly in the case of chickpea and pigeon pea. For technology development to improve pulse yield, it must encompass some unique features, such as the need to fit pulse cultivation into the cereal-farming complex in crop rotation and intercropping, which makes attributes like crop duration very important. It also has to deliver productivity in marginal environments where irrigation is lacking and farmers have little purchasing power. The successful development of short-duration and wilt-resistant chickpea varieties, for example, has led to their adoption in new niches in southern India and in the rainfed rice fallow

FIGURE 1.2A Pulse yield at aggregate level in India, 1961–2013

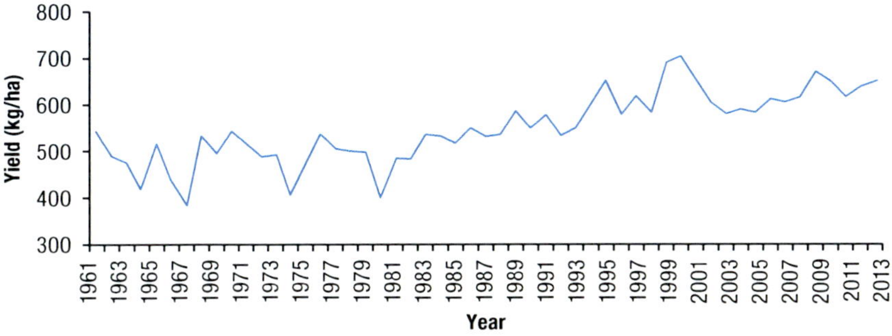

Source: India, Ministry of Agriculture, *Agricultural Statistics at a Glance*, various years (accessed May 2015).

FIGURE 1.2B Yield by pulse type in India, 1961–2013

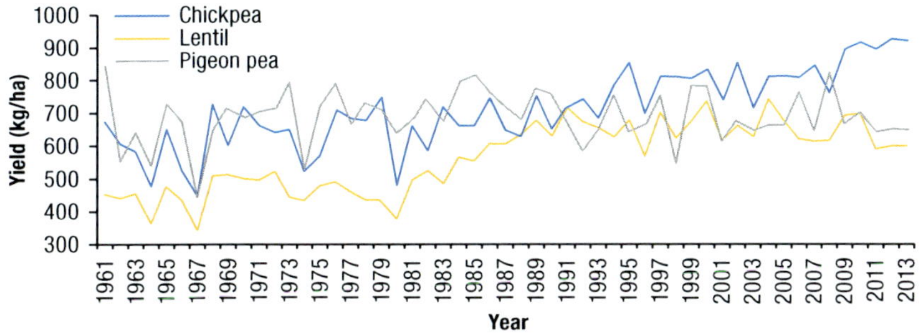

Source: India, Ministry of Agriculture, *Agricultural Statistics at a Glance*, various years (accessed May 2015).

lands (Gowda and Gaur 2004; Gaur et al. 2008). This book aims to draw useful lessons from these successes.

Figure 1.2a and Figure 1.2b show that the little yield growth that has happened is concentrated in few pulses, mainly chickpea. Even at its peak production, chickpea yield was much lower in India than in leading countries like Israel (3 tons per hectare), Australia (more than 2 tons), and China (2 tons). Similarly, for pigeon pea, countries like the Philippines attain yields greater than 1 ton per hectare compared with a peak yield of about 800 kilograms per hectare in India. In 2013, pigeon pea yields were 650 (kilograms per hectare) for India; 2,520 for Canada; 2,037 for the United States; 1,550 for China; and 1,409 for Australia (FAO 2013).

Global Trade and India's Performance over Time

Among the other motivations justifying the need for this book is the need to better understand the potential and actual effects of India's pulse sector on other countries and the effects of other countries' production on India. Figure 1.3 shows the shifting rank in production of the top global producers since 1961. While other countries have switched ranks (for example, China), India has consistently remained the largest producer. The Indian trade deficit in pulses, however, has turned some countries into large producers and sellers as they have found a big market opportunity in Indian consumers. For example, Myanmar, which before 2000 was not ranked among the world's top five producers, now ranks as the largest exporter of pigeon pea, black matpe, and green gram to India. India's pulse market has also had significant spillover effects for countries as far away as Canada and Australia, and most recently for several African countries, including Malawi, Mozambique, and Tanzania. The availability of African pigeon pea production is synchronous with the seasonal incidence of high prices in the Indian market, since the bulk of African pigeon pea exports occur from September to January, before the harvest of India's own rainy-season crop (Walker et al. 2015).

The advent of imports of pulses by India is a relatively new phenomenon. Until the late-1990s, India continued to be nearly self-reliant in this product and did not require any sizable imports, but as Figure 1.4 shows, the country is now the largest importer of pulses, with imports making up as much as a quarter of its total pulse consumption in some years. Before 2001, India figured among the top five importers only once, in the 1980s. India's pulse imports are concentrated across just four or five countries, which generally constitute up to 95 percent of what it imports. Since trade has grown significantly (the import values equal nearly US$1.5 billion) and has become a sizable part of the consumption portfolio, this book examines the dynamics of pulses trade in detail. Currently, pulse imports are second only to edible oils in terms of import penetration in food.

Nutritional Value and Cost to Consumers

This book is also motivated by the role of pulses in nutrition. Pulses are important components of the Indian diet and constitute a major source of quality protein. They provide other nutrients, including carbohydrates, dietary fiber, unsaturated fat, and vitamins and minerals, as well as non-nutrients, such as antioxidants and phytoestrogens. Their most common and widely recognized role is in supplying proteins for vegetarian consumers. The

FIGURE 1.3 Top producers of pulses in the world, 1961–2011 (in millions of metric tons)

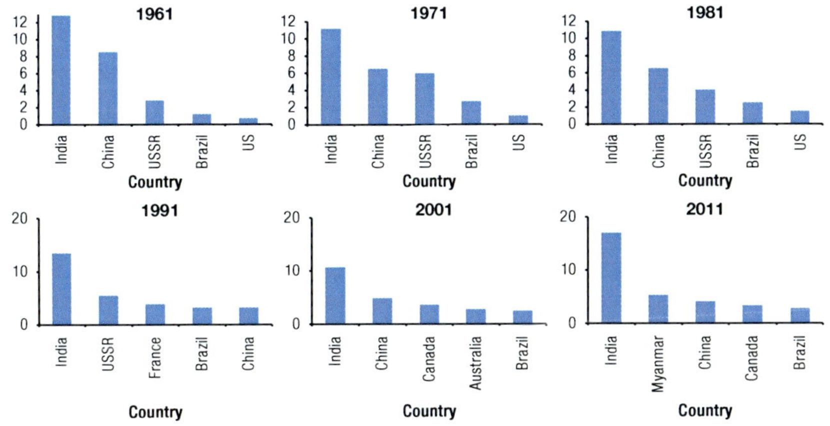

Source: FAO (2013).

FIGURE 1.4 Top importers of pulses in the world, 1961–2011 (in millions of metric tons)

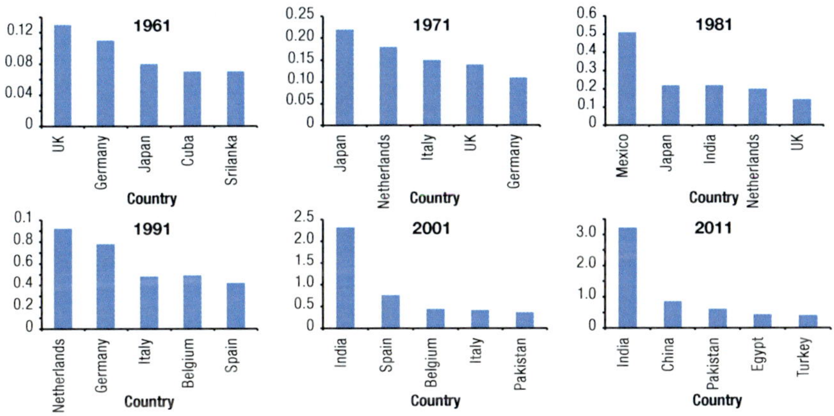

Source: FAO (2013).

latest round of the NSS (68th round) found that in rural India pulses contribute 10 percent of protein intake, the same proportion as milk and milk products contribute, and in urban India slightly more (11 percent). In some states, pulses contribute as much as 14 percent of protein intake, and in many states, they contribute more than milk or meat-fish-eggs. In India's urban sector the contribution of pulses to protein was in the range of 10 to 13 percent for 13 of the major states.

The value of pulses as a source of protein is important today because, among both poor and middle-income households, protein intake levels have declined since 1988. Indeed, based on the NSS data over different rounds, in India's rural areas this fall in protein intake has even affected rich households. In short, the only population in the country whose protein intake has improved over the past two decades is that of urban rich households. All of this places pulses in a comparatively important role, particularly in rural India, where a 2010 study (Arlappa et al. 2010) found that 73 percent of households did not consume the recommended dietary intake (RDI) of pulses. The latest NSSO survey indicates that 50 percent of the Indian population consumes less than the recommended daily intake of protein, which is 60 grams a day (NSSO 68 report 560, Table T11, NSSO 2014). The deficiency is more pronounced in the lower income strata (for example, the bottom 20 percent of the population consumes only 80 percent of the recommended protein intake).

The rising cost of food is certainly relevant to the decline in protein intake. Mishra and Roy (2011) examined the drivers of food inflation in India and found that the rate of price increases after 2005 has been high in pulses, although it has been relatively higher in milk and milk products as well as fish and meat. Two pulses, particularly pigeon pea and green gram, have consistently been in the cluster of pulses undergoing high price increases.

The latest round of NSS data provides information on nutrients per unit of food, and Table 1.2 lists, for comparison, these measures for calories, protein, and fat for a number of the most common foods alongside seven pulses. It is evident that per unit of weight, pulses provide high levels of protein. Whether they turn out to be the cheapest source of protein for a household varies across time.

Government Policy Responses

As discussed earlier, the performance of the pulses sector in India seems to have been subpar, with consistent gaps between demand and supply over a long period leading to more than a billion dollars' worth of annual imports. Importantly, this has happened despite many large-scale government initiatives combined with such policies as a hike in the MSP (even though there is limited procurement) to stimulate the pulses sector. For example, the government has launched several productivity enhancement programs. In 1967 it established the All India Coordinated Pulses Improvement Project (AICPIP), which was later elevated to the Indian Institute of Pulses Research (IIPR).[6]

6 Pulses are not unique in this context in terms of a coordinated research program.

TABLE 1.2 Nutrients per kilogram for common food items in India

Item	Unit	Calories per unit (kilocalorie)	Protein per unit (grams)	Fat per unit (grams)
Rice	kilograms	3,460	75	5
Wheat	kilograms	3,410	121	7
Lowar (sorghum) and products	kilograms	3,490	104	19
Bajra (pearl millet) and products	kilograms	3,032	97	42
Maize and products	kilograms	3,420	111	36
Pigeon pea	kilograms	3,350	223	17
Chickpea (split)	kilograms	3,720	208	56
Chickpea (whole)	kilograms	3,720	208	56
Green gram	kilograms	3,480	245	12
Lentil	kilograms	3,430	251	7
Black matpe	kilograms	3,470	240	14
Peas	kilograms	3,150	197	11
Other pulses	kilograms	3,400	220	12
Milk	liters	1,000	40	70
Eggs	number	100	8	8
Fish, prawn	kilograms	1,050	140	20
Goat meat	kilograms	1,180	214	36
Chicken	kilograms	1,090	259	6

Source: NSSO 2014 and Indian Council of Medical Research (ICMR).

Pulses also have received significant attention in different five-year plans, including the Intensive Pulses District Program launched during the Fourth Five-Year Plan (1969–1974), the National Pulses Development Program launched during the Seventh Five-Year Plan (1985–1989), and a special food grain production program launched in 1988–1989.

In 2004 schemes for pulses, along with schemes for oilseeds, oil palm, and maize, were brought under one centrally sponsored scheme: the Integrated Scheme of Oilseeds, Pulses, Oilpalm, and Maize (ISOPOM) (India, Ministry of Agriculture, *Agricultural Statistics at a Glance*).[7] In 2007 pulses were also made a focus crop in the National Food Security Mission (NFSM) in 171 districts across 14 states, and in 2011–2012 the government allocated

7 The schemes were: ODP (Oilseed Development Programme), OPDP (Oil Palm Development Programme), NPDP (National Pulses Development Programme), and AMDP (Accelerated Maize Development Programme).

3 billion rupees for the integrated development of 60,000 pulse villages under Rashtriya Krishi Vikas Yojana (RKVY) (the national agricultural development plan). The government of India has also followed a liberal trade policy in pulses. Many of the trade barriers that were in place before the 1970s were removed to encourage cheap imports for general pulse consumption.

Despite the wide range of research and several government programs and policy stances, India's pulses sector has recorded barely any growth either in the area planted or in yield for the past five decades. From the point of view of this volume's contributors, the stubborn lack of change on the ground despite all those efforts represents one of the compelling reasons this important sector needs to be studied more closely. Rigorous impact assessments of the government's several initiatives have not been done, apart from a summary evaluation by the government concerning the NFSM (India, Ministry of Agriculture 2014). In that evaluation, the government reported that after launching the mission in 2007–2008, the area covered under pulses increased by 3.1 percent but then declined during 2008–2009 by 4.8 percent. Pulse cultivation again picked up momentum in 2009–2010, and it registered a growth rate of 29 percent at the all-India level. Another evaluation of the NFSM, by Thomas, Sundaramoorty, and Jha (2013), found that there were significant increases in pulse production with increases in area in Andhra Pradesh, Karnataka, and Maharashtra but significant increases in yields in just 2 of the mission's 14 states (Karnataka and Maharashtra).

How This Book Is Organized

Chapter 2 examines the state of pulse demand and its distribution across space and over time. Among the findings, the chapter shows that the fall in per capita consumption of pulses has been consistent across all household income groups and across both rural and urban regions. Trends over time in pulse consumption are described in detail across different income and demographic groups, including projections for pulse demand extending to 2030. The chapter, by outlining the nutritional contribution of pulses, suggests that enhancing consumers' access to pulses could improve nutritional status in terms of protein intake. High price responsiveness, particularly among the poor, suggests that managing inflation in pulse prices is likely to be important for raising pulse consumption.

Chapter 3 gives a detailed description of pulse production dynamics across regions and over time. It divides the history since 1960 into four time periods: pre- and early Green Revolution; advanced and post–Green Revolution;

postliberalization; and post–trade spike. To study the spatial movement in pulse cultivation, states are grouped into five geographic zones: north, east, south, west, and central. The chapter formally establishes that pulse production has shifted from traditional to nontraditional areas over time, moving from north to south and from east to west and central regions, with Madhya Pradesh and Chhattisgarh becoming the hub of pulses production.

Chapter 4 looks at the technology, with special importance given the stagnation in pulses' yield as compared with that of other crops. Yield improvement in pulses will be needed to overcome the position of advantage that now favors cereals and oilseeds. To date, the history of technology improvements in pulses has been mixed, with the pace of development and adoption picking up only in the past decade or so. With muted supply response in pulses driven by technology and agricultural policies, India has faced a persistent deficit in pulses that has led to significant imports.

Chapter 5 studies another important segment of the pulses sector: food processing. Given that NSS data show an increase in consumption of processed food items, declining per capita consumption of pulses could possibly be checked by the development of innovative pulse products that processing could bring about. However, the report card for this sector is far from encouraging, as much pulse processing continues to be done with low levels of technology and at relatively small scale. The few examples of pulse processing that are claimed to be successful nevertheless suggest that introducing an element of product differentiation might offer promise to the sector, particularly in marketing pulses as a health food.

Chapter 6 addresses the problem of persistent and rising reliance on pulse imports. It looks at the expansion of trade along both the intensive and extensive margins. With the help of liberal trade policies—low import tariffs and export restrictions—the government tried to enhance the availability of pulses in the country. The chapter shows that, especially throughout the 2000s, pulse imports rose significantly, with import penetration increasing from 10 percent to 20 percent. It also notes how significant changes have taken place in the import basket. Whereas chickpea and pigeon pea had dominated the import basket in the first half of the 2000s, in the latter half, they were replaced by yellow pea, a variety not even produced in India. Yellow pea's consistent importation from Canada in significant quantities and at low prices has created a sizable demand for it. These changes allude to the possibilities of the roles played by trade and the potential risks involved (if, for example, trade largely expands on intensive margin). The chapter also documents new pulse exports emanating from African countries, including Tanzania, where some Indian firms are beginning to lease-in land to export pulses to India. The effect of imports on domestic prices for one of India's most important pulses (in production and consumption)—pigeon pea—are also analyzed. Results are nuanced, in the sense that the dampening effect of imports on prices has not been large enough to actually bring prices down but has instead arrested the rate of price increase.

Chapter 7 reviews the role of convergent innovation for the development of the pulses sector and discusses an evolving framework focused on the significant health and environmental benefits that could accrue from pulses. Case studies of convergent innovation have shown improved outcomes that this approach can make possible, and it seems that the use of this approach to study the prospects of India's pulses sector could be useful. Finally, Chapter 8 concludes, drawing lessons from the studies in the preceding chapters to provide policy suggestions for the way forward.

References

Arlappa, N., A. Laxmaiah, N. Balakrishna, and G. N. V. Brahmam. 2010. "Consumption Pattern of Pulses, Vegetables, and Nutrients among Rural Population in India." *African Journal of Food Science* 4 (10): 668–675.

Banerjee, G., and L. M. Palke. 2010. *Economics of Pulses Production and Processing of India*. National Bank for Agriculture and Rural Development (NABARD) Occasional Paper 51. Mumbai, India.

Boye J., F. Zare, and A. Pletch. 2010. "Pulse Proteins: Processing, Characterization, Functional Properties, and Applications in Food and Feed." *Food Research International* 43 (2): 414–431.

Census of India. 2014. Baseline Survey 2014. Sample registration system. Accessed January 2015. www.censusindia.gov.in/vital_statistics/baseline%20TABLES07062016.pdf.

Dwivedi A., I. Dev, V. Kumar, R. S. Yadav, M. Yadav, D. Gupta, A. Singh, and S. S. Tomar. 2015. "Potential Role of Maize-Legume Intercropping Systems to Improve Soil Fertility Status under Smallholder Farming Systems for Sustainable Agriculture in India." *International Journal of Life Sciences Biotechnology and Pharma Research* 4 (3): 145–157.

FAO (Food and Agriculture Organization of United Nations). 2011, 2012, and 2013. FAOSTAT database. Accessed in April 2015 and July 2015. http://faostat3.fao.org/home/E.

Gaur, P. M., J. Kumar, C. L. L. Gowda, S. Pande, K. H. M. Siddique, T. N. Khan, T. D. Warkentin, S. K. Chaturvedi, A. M. Than, and D. Ketema. 2008. "Breeding Chickpea for Early Phenology: Perspectives, Progress and Prospects." In *Food Legumes for Nutritional Security and Sustainable Agriculture*, vol. 2, edited by M. C. Kharkwal, 39–48. New Delhi: Indian Society of Genetics and Plant Breeding.

Gokarn, S. 2011. "The Price of Protein." *Macroeconomics and Finance in Emerging Market Economies* 4 (2): 327–335.

Gowda, C. L. L., and P. M. Gaur. 2004. "Global Scenario of Chickpea Research: Present Status and Future Thrusts." In *Pulses in New Perspective*, edited by Masood Ali, B. B. Singh, Shiv Kumar, and Vishwa Dhar, 1–22. Kanpur, India: Indian Society of Pulses Research and Development.

Gowda C. L. L, S. Srinivasan, P. M. Gaur, and K. B. Saxena. 2013. "Enhancing the Productivity and Production of Pulses in India." In *Climate Change and Sustainable Food Security*, edited by P. K. Shetty, S. Ayyappan, and M. S. Swaminathan, 145–159. Bangalore: National Institute of Advanced Studies; New Delhi: Indian Institute of Pulses Research (IIPR).

ICMR (Indian Council of Medical Research. 2009). "Nutrient Requirements and Recommended Dietary Allowances for Indians: A Report of the Expert Group of the Indian Council of Medical Research." New Delhi.

IIPS (International Institute for Population Sciences) and ORC (Opinion Research Corporation) Macro. 2007. *National Family Health Survey (NFHS-3), 2005–06: India*. Mumbai, India: International Institute for Population Sciences.

India, Ministry of Agriculture. Various years. *Agricultural Statistics at a Glance*. Department of Agriculture and Co-operation. Accessed May 2015.

———. 2014. "Report on the Impact Evaluation of National Food Security Mission [NFSM]." Accessed July 2015. www.nfsm.gov.in/Publicity/2014-15/Books/NFSM%20IE-FINAL%20 REPORT-231114.pdf.

Jukanti, A. K., P. M. Gaur, C. L. L. Gowda, and R. N. Chibbar. 2012. "Nutritional Quality and Health Benefits of Chickpea (*Cicer arietinum L.*): A Review." *British Journal of Nutrition* 108: S12–S26.

McCrory M. A., B. R. Hamaker, J. C. Lovejoy, and P. E. Eichelsdoerfer. 2010. "Pulse Consumption, Satiety, and Weight Management." *Advances in Nutrition* 1: 17–30.

Mishra, P., and D. Roy. 2011. "Explaining Inflation in India: The Role of Food Prices." Paper presented at the India Policy Forum, National Council of Applied Economic Research, New Delhi, October 31. www.prachimishra.net/mishra_roy_reviseddraft_ipf_oct25.pdf.

Mohanty, S., and K. J. Satyasai. 2015. "Feeling the Pulse, Indian Pulses Sector." NABARD Rural Pulse, Issue X (July–August): 1–4.

NCAER (National Council of Applied Economics Research). 2014. "India's Pulses Scenario." A research paper prepared under the project Agricultural Outlook and Situation Analysis Reports. Available at http://nfsm.gov.in/Meetings_Minutes/NCAER/ResearchPapersNCAER/7%20India's%20pulses%20Scenerio.docx. Accessed March 2016.

NSSO (National Sample Survey Office). Various years. Consumption Expenditure Surveys. New Delhi: Ministry of Statistics.

———. 2014. *Household Consumption of Various Goods and Services in India, 2011–2012*. NSSO 68th Round. Ministry of Statistics and Program Implementation, Report 558. Accessed June 2016. http://mospi.nic.in/mospi_new/upload/Report_no558_rou68_30june14.pdf.

Pande, S. P., and P. K. Joshi. 1995. *Constraints and Prospects of Legumes in Northern India and the Tarai Region of Nepal*. Patancheru, India: International Crops Institute for Semi-Arid Tropics.

Peoples, M. B., J. Brockwell, D. F. Herridge, I. J. Rochester, B. J. R. Alves, S. Urquiaga, R. M. Boddey, F. D. Dakora, S. Bhattarai, S. L. Maskey, C. Sampet, B. Rerkasem, D. F. Khans, H. Hauggaard-Nielsen, and B. S. Jensen. 2009. "The Contributions of Nitrogen-Fixing Crop Legumes to the Productivity of Agricultural Systems." *Symbiosis* 48: 1–17.

Rammohan, A., N. Awofeso, and M.-C. Robitaille. 2012. "Addressing Female Iron-Deficiency Anaemia in India: Is Vegetarianism the Major Obstacle?" *ISRN Public Health*, Article ID 765476, 8 pages, doi:10.5402/2012/765476.

Reddy A. A. 2004. "Consumption Pattern, Trade, and Production Potential of Pulses." *Economic and Political Weekly* 39 (44): 4854–4860.

Reddy, A. A., M. C. S. Bantilan, and G. Mohan. 2013. *Pulses Production Scenario: Policy and Technological Options*. Policy Brief 26. Patancheru, India: International Crops Institute for Semi-Arid Tropics.

Reddy, A. A., and G. P. Reddy. 2010. "Supply Side Constrains in Production of Pulses in India: A Case Study of Lentil." *Agricultural Economics Research Review* 23 (January–June): 129–136.

Sonna T., H. Joshi, A. Sebastian, and U. Sharma. 2014. "Analytics of Food Inflation in India." WPS (DEPR): 10 / 2014. RBI Working Paper Series, Department of Economic and Policy Research. October.

Srivastava, S. K., N. Sivaramane, and V. C. Mathur. 2010. "Diagnosis of Pulses Performance of India." *Agricultural Economics Research Review* 23: 137–148.

Thomas L., C. Sundaramoorty, and G. Jha. 2013. "The Impact of National Food Security Mission on Pulse Production Scenario in India: An Empirical Analysis." *Indian Journal of Agricultural Statistics* 9 (1): 213–223.

Tiwari, B., A. Gowen, and B. Mckenna. 2011. *Pulses Foods Processing, Quality, and Nutraceutical Applications*. Elsevier e-book, first edition. London NW1 7BY, UK.

Walker, T., S. Silim, B. Cunguara, C. Donovan, P. R. Parthasarathy, M. Amane, and M. Siambi. 2015. "Pigeonpea in Mozambique: An Emerging Success Story of Crop Expansion in Smallholder Agriculture." Report. Modernizing Extension and Advisory Services project, University of Illinois at Urbana-Champaign, Illinois, US.

Chapter 2

CHANGING CONSUMPTION PATTERNS AND ROLES OF PULSES IN NUTRITION, AND FUTURE DEMAND PROJECTIONS

Praduman Kumar, P. K. Joshi, and Shinoj Parappurathu

In this chapter we study the dynamics of consumption of pulses and assess pulses' nutritional role with a focus on proteins. We note variations among different income strata, highlighting the higher protein deficiency among low-income groups and the higher price responsiveness of poorer consumers. We discuss the likely ineffectiveness of targeted consumer subsidies. Subsequently, using demand-system estimation, we make projections for the demand for pulses in India to 2030.

Background

The consumption of pulses per capita in India has been persistently below the recommended levels. The current average rate of consumption—47 grams per capita per day—is marginally higher than what is recommended for people with a sedentary lifestyle by the Indian Council of Medical Research (ICMR) (40 grams per capita per day), but it is much lower than the recommendation for working men and women (60 and 50 grams, respectively). Pulses are an important source of protein in Indian diets. Based on the latest round of NSS (National Sample Survey) data and nutrition charts from the Indian Council of Medical Research, pulses tend to be among the cheapest sources of protein, despite their persistently rising prices over the past decade. Table 2.1 shows that cereals contribute the highest amounts to protein availability, but pulses are among the cheapest sources of protein, and peas are the cheapest among pulses.

The animal source foods (ASF) are the most expensive sources of protein in India. In a famous study, Patwardhan (1962) elaborated on the role legumes play in the diets of populations in the tropics and subtropics in contexts where ASF consumption is less common. In these areas the use of comparatively protein-rich legumes is an essential strategy in people's attempt to balance their diets. The seeds of pulses contain two to three times more protein

TABLE 2.1 Protein contribution and its cost across food (rural and urban)

	66th round (rupees per kilogram of protein)		68th round (rupees per kilogram of protein)		
Food item	Rural	Urban	Rural	Urban	Protein per unit (grams)
Pulses					
Pigeon pea	317	341	260	290	223
Gram, split	182	197	217	232	208
Gram, whole	173	203	200	237	208
Green gram	272	288	259	284	245
Lentils	244	254	208	223	251
Black matpe	236	263	235	259	240
Peas	134	166	154	192	197
Gram flour	185	194	217	226	220
Animal source foods					
Milk (liter)	462	558	613	731	40
Egg (number)	387	382	447	437	8
Fish/prawn	499	622	611	758	140
Goat/mutton	823	907	1,094	1,220	226
Beef/buffalo	392	405	490	478	214
Chicken	397	412	447	463	259
Cereals					
Rice, PDS	47	41	61	79	75
Rice, market	223	301	247	320	75
Wheat, PDS	41	53	40	52	121
Wheat, other sources	102	125	102	132	121
Maize			105	225	111
Coarse cereals					
Pearl millet	109	64	115	144	97
Sorghum	113	264	214	254	104
Finger millet	148	180	182	201	73
Other millets	137	515	227	515	97
Other cereals	258	747	280	427	97

Source: NSSO (2014a); NSSO (2014b).

than cereals and root tubers (Table 2.1). Depending on the species and variety, pulses have a protein content that ranges between 17 percent and 32 percent. Those most commonly consumed—chickpea, pigeon pea, and black matpe—are among the richest sources of proteins among vegetarian food items. Pulses can also improve the protein intake of meals in which cereals and root tubers are combined with them (Khushwaha, Rajawat, and Kushwah 2002). More than 40 years ago, Sukhatme (1970) showed that diets based on cereals and pulses normally consumed in India could meet the needs for protein at all ages provided that enough food is taken to satisfy energy needs.

Besides being rich in protein, pulses contain a wide range of nutrients, including carbohydrates, dietary fiber, unsaturated fat, vitamins, and minerals, as well as non-nutrients, such as antioxidants and phytoestrogens. Pulses contribute to reduced colon cancer and cardiovascular disease, increased satiety, and lowered Body Mass Index and obesity risk (Boye, Zare, and Pletch 2010; McCrory et al. 2010; Jukanti et al. 2012). These health benefits assume great importance in the wake of increasing incidence of diabetes and other lifestyle-related diseases. Such diseases are expected to account for a worldwide healthcare burden of US$47 trillion from 2010 to 2030 (Alwan et al. 2010).

In India, the importance of pulses is also driven by the high incidence of vegetarianism. Although nonvegetarian foods like fish and other seafood, meat, and eggs are better sources of quality proteins, they are not consumed by the vegetarian consumers who constitute a substantial portion (28 to 29 percent) of the Indian population. These animal sources of protein tend to be expensive, so their regular availability is limited among the many consumers whose purchasing power is weak (Table 2.1). Supplementing cereal-based diets with pulses is therefore regarded as a potential solution for reducing India's high level of protein-energy malnutrition. To achieve maximum supplementary effects, the ICMR recommends the following combination: four parts cereal protein + one part pulse protein. In terms of grains, this formula means eight parts cereals + one part pulses (Gopalan, Rama Sastri, and Balasubramanian 1989).

Furthermore, from a nutritional standpoint the particular amino acid composition of pulse proteins is very important. The nutritive value of a protein or a protein mixture for various functions in the animal body is related to the proportions of essential amino acids contained in it. Vijayaraghavan and Srinivasan (1953) studied the values for the essential amino acid composition of five Indian pulses: chickpea, pigeon pea, green gram, cowpea, and lentil. Their results, which express the essential amino acid contents per 100 grams of flour, showed that as compared with whole eggs, pulses' composition included certain amino acids, such as tryptophan and methionine-cysteine, that are limiting amino acids, and that these limiting amino acids vary from one type of pulse to another.[1]

Proteins are made of amino acids, nine of which are essential and need to be obtained from the diet to prevent protein-energy malnutrition. Pulses (and animal products) have a relatively high content of the essential amino acid lysine, while cereals have considerably less of it. Pulses, however, have less of the essential amino acids methionine and cysteine, while cereals have more of them. Nutritionally, this implies that the inclusion of enough different pulses combined with rice or wheat can be of high supplementary value in supplying essential amino acids to the individual (Vijayaraghavan and Srinivasan 1953).

Study Objectives and Data Sources

With this background, next we look at the changing patterns of pulse consumption in India as well as the nutritional role pulses play, particularly as a provider of proteins. The objectives of this analysis are (1) to identify the historical trends in the consumption of pulses and their products across income groups, demographic areas (rural and urban), and states; (2) to examine the changing patterns in pulses' contribution to households' dietary protein, across income groups and demographic areas; (3) to analyze the statewide changing trends in protein intake from pulses in India over time; and (4) to project the demand for pulses as of 2020 and 2030.

The study relies on unit-level data on dietary patterns and consumer expenditure collected by the National Sample Survey Organization (NSSO) under its 43rd (1988–1989), 55th (1999–2000), 66th (2009–2010), and 68th (2011–2012) rounds. It was felt in several quarters that the 2009–2010 period could be an abnormal year, given both the global financial slowdown then hitting urban areas and a drought affecting rural India, so the 68th round was done in 2011–2012, only a year after the 66th round of NSSO surveys. Treating per capita expenditure as a proxy for per capita income, we categorized the sample households into three expenditure/income groups: (1) poor (below poverty line); (2) middle income (between 100 percent and 150 percent of the poverty line); and (3) rich (above 150 percent of the poverty line). In some cases, the classification involved two broad groups—that is, poor and

1 Limiting amino acids refer to those amino acids that limit the functioning of the proteins.

nonpoor. The analysis was carried out both by income groups and by demographic (rural and urban) areas for different states of India.

Dynamics of Pulses Consumption

In India, per capita pulse consumption has declined considerably from 1988–1989 to 2009–2010, falling by 27.2 percent over that period, placing it second in reduction only after coarse grains among the different food categories measured. During this entire period, consumption of pulses and pulse products rose only once, between 2004–2005 and 2011–2012, although it rose again in the years 2013–2014 and 2014–2015. As of 2011–2012, four pulses—pigeon pea, green gram, lentil, and black matpe—together made up about 64 percent of the consumption of pulses and pulse products in rural India and 68 percent in urban India. Pigeon pea accounted for as much as 27 percent of pulse consumption in rural areas and 33 percent in urban areas. Green gram and lentil together contributed 26 percent in rural areas and 23 percent in urban areas, with the share of green gram being greater in urban India. Split chickpea contributed about 10 percent in each sector. Products of pulses and chickpea had a total share of 9 percent in rural areas and 11 percent in urban areas.

Between 2004–2005 and 2011–2012, consumption of pulses and pulse products rose by 77 grams to 78 grams (NSS 68th round report), increasing about equally in rural areas (from 705 grams to 783 grams per capita per month) and urban areas (from 824 grams to 901 grams). The majority of this rise—accounting for as much as 69 grams in the rural sector and 57 grams in the urban sector—consisted of increases in the consumption of just four items: split chickpea, whole chickpea, pea, and chickpea flour (a processed pulse). Hence, based on the last three rounds of NSS data, over the seven-year period beginning in 2004, these pulses registered only a modest increase in monthly per capita consumption, specifically a 14 gram increase in the rural sector and an 18 gram increase in the urban sector (NSSO 2014a).

Table 2.2 presents the dynamics of pulses consumption in detail, showing changes in per capita consumption across different household income categories over approximate 10-year intervals (for 1988–1989, 1999–2000, 2009–2010, and 2011–2012) for both rural and urban areas. Pulse consumption in India declined from 11.6 kilograms (per capita per year) in 1988–1989 to 10.5 kilograms by 1999–2000 and then declined further to 8.5 kilograms by 2009–2010. With a modest turnaround, based on the latest round of NSS data in 2011–2012, per capita consumption rose to 9.6 kilograms in rural areas and 10.5 kilograms in urban areas. The data also reveal that the rate of

TABLE 2.2 Dynamics of pulses consumption across households of different income groups in rural and urban India, 1988–2011 (in kilograms per capita per year)

Income group of households	1988	1999	2011	Change (%) (1988–2011)	Annual growth (%)
Rural India					
Poor	7.4	6.3	6.9	−6.8	−0.3
Middle income	10.6	9.8	8.6	−18.9	−0.9
Rich	15.6	12.8	11.5	−26.3	−1.3
All	11.2	10.0	9.6	−14.3	−0.7
Urban India					
Poor	8.7	7.8	7.3	−16.1	−0.8
Middle income	11.8	12.0	8.9	−24.6	−1.2
Rich	16.5	18.4	12.3	−25.4	−1.3
All	12.5	14.5	10.5	−16.0	−0.8
India					
Poor	7.9	6.9	7.1	−10.1	−0.5
Middle income	11.0	9.3	8.7	−20.9	−1.0
Rich	15.8	13.2	11.9	−24.7	−1.2
All	11.6	10.5	10.0	−13.8	−0.6

Source: Authors' calculations based on NSS data.

Note: Poor = below poverty line; middle income = between 100 percent and 150 percent of poverty line; rich = above 150 percent of poverty line. The All India figures from NSS report 560 for the latest round 68 are used and cover the major states only.

decline in pulse consumption has increased with household income level, particularly in rural India. For example, between 1988–1989 and 2011–2012, the per capita consumption of pulses dropped from 7.9 kilograms to 7.1 kilograms among poor households, from 11 kilograms to 8.7 kilograms among middle-income households, and from 15.8 kilograms to 11.9 kilograms among rich households. In percentage terms, the decline was much greater for rich (24.7 percent) and middle-income (20.9 percent) households than for poor ones (10.1 percent). This translates to annual reductions at the rates of 1.23 percent, 1.01 percent, and 0.46 percent, respectively.

Moreover, an urban-rural comparison across the same income group shows a greater decline in pulse consumption among the urban poor (by 16.1 percent) than the rural poor (6.8 percent), while the rate of decline was slightly lower among the urban rich (25.4 percent) than the rural rich (26.3 percent) until 2010–2011. In general, the trends seem to show a convergence toward bridging the gaps in consumption across different income groups as well as across rural-urban categories.

The changes in pulse consumption across all of India's states are presented in Tables 2A.1 and 2A.2 in the chapter appendix. The consumption levels in 1988–1989, 2009–2010, and 2011–2012, presented in Table 2A.1, show a declining trend (though to varying degrees) across both rural and urban households in all states except Kerala until 2009–2010. In particular, the consumption dropped by 30 percent or more across rural households in the states of Bihar, Delhi, Jharkhand, Madhya Pradesh, Uttar Pradesh, and most northeast states, and across urban households in the states of Andhra Pradesh, Assam, Bihar, Jharkhand, Kerala, Rajasthan, and Uttarakhand besides Delhi and the northeast states.

Table 2A.2 reveals a decline of 20 percent or more across rich as well as poor households in all the states, although the decline has been steeper among rich households than poor ones. As discussed earlier, the latest round of NSS data (for 2011–2012) shows that the consumption of pulses has recently experienced modest increases in most states, although in some states there has been almost no change. In no states have the consumption levels returned to where they were in 1988–1989, and in some cases they are not as high as in 2004–2005. This comparison is pertinent because 2009–2010 consumption levels could have been affected to some extent by droughts. Yet the results are quite informative for gauging the consumption effects on pulses during periods of shock, which do happen from time to time given the rainfed conditions under which pulses are grown in India (see Chapter 3 for more on this subject). The turnaround in consumption in 2011–2012 should also be looked at in relation to the shock period of 2009–2010.

Pulse prices have been rising consistently for quite some time (only recently they have moderated), resulting in further decline in pulse consumption from an already low level. An average Indian consumed 60 grams of pulses per day in the 1950s. Data from consumption expenditure surveys conducted by the National Sample Survey Organization show that the rise in retail price is a major reason for their declining consumption. Part of the reason for the high prices of pulses has been that the production and productivity of pulses have registered very slow growth in India over the last five decades. Pulse production remained stagnant at around 14 million metric tons annually for decades, from the 1950s to the early years of the twenty-first century, before it increased to 17 million to 18 million metric tons in 2013–2014, where production has hovered ever since.[2] The increase in production has been slow in other parts of the world too (Rao and Joshi 2016). Thus the availability of

2 Hereafter "tons" refers to "metric tons."

pulses in India as well as in the global markets has not kept pace with the rising demand. Moreover, the recent increases in pulse production have often been reversed by repeated droughts in large parts of India in the past two years (2013 and 2014). Apart from rising prices, there has also been a secular shift away from cereals and pulses toward high-value products like fruits and vegetables and animal source foods and processed items owing to changing preferences driven by such factors as rising incomes, urbanization, and greater participation of women in the labor force.

Analysis of Pulse Consumption at a Disaggregated Level

The major pulses consumed in India are chickpea, pigeon pea, green gram, black matpe, lentil, and peas. Among these, chickpea and pigeon pea together account for nearly half of consumption, with pigeon pea ranking highest in quantity consumed among both poor and rich households until the latest round of the NSS data. Both of these two pulses, however, experienced a considerable decline in consumption levels during the past two decades; in urban areas, chickpea declined by 78.4 percent and pigeon pea by 7.7 percent between 1988–1989 and 2011–2012 (Table 2.3). Over the same period, urban consumption of green gram, lentil, and black matpe also fell—by 26.3 percent, 31.3 percent, and 15.4 percent, respectively. Although the general trend of declining consumption has been pervasive across all major pulses, some cheaper pulses, like peas, have experienced a considerable rise in consumption levels. The most recent NSS data from the 68th round show a modest turnaround in most pulses in both rural and urban areas, but as discussed earlier, the levels have not yet reached the high of two decades ago or that reached in 2004–2005 (Table 2.3).

Furthermore, among pulses, the consumption of chickpea has remained stable across poor households over the same period (1988–2009), but it has declined among rich households (Table 2.4). The consumption of green gram, lentil, and black matpe has also dropped considerably, somewhat more steeply among the poor than among the rich. Countering these trends has been the consumption of peas, which while still among the least consumed among the major pulses, rose almost twofold in consumption among the poor during this period, probably because of its increased availability, particularly as low-priced imports (see Chapter 6 for further discussion on this topic). The numbers in Table 2.4 present the figures up to the 66th round of the NSS (2008–2009).

TABLE 2.3 Trends in consumption levels of major pulses in rural and urban India, 1988 to 2011–2012 (in kilograms per capita per year)

	Rural			Urban		
Pulses	1988–1989	2009–2010	2011–2012	1988–1989	2009–2010	2011–2012
Chickpea	2.4	0.9	0.9	2.5	0.5	0.5
Pigeon pea	2.8	1.9	2.6	3.9	2.6	3.6
Green gram	1.4	0.9	1.0	1.9	1.2	1.4
Lentil	1.7	1.2	1.3	1.6	1.2	1.1
Black matpe	1.2	0.9	1.0	1.3	1.0	1.1
Peas	0.3	0.5	0.6	0.3	0.4	0.2
Total pulses	11.2	8.1	9.3	12.4	9.0	10.8

Source: Authors' calculation based on NSS data, various years.

TABLE 2.4 Trends in consumption of major pulses across poor and rich households in India, 1988–2009 (in kilograms per capita per year)

	Poor households			Rich households		
Pulse	1988	2009	Change (%)	1988	2009	Change (%)
Chickpea	1.4	1.4	0.0	3.7	2.6	−29.7
Pigeon pea	2.2	1.2	−45.5	4.2	2.8	−33.3
Green gram	0.9	0.5	−44.4	2.3	1.3	−43.5
Lentil	1.3	0.8	−38.5	2.1	1.4	−33.3
Black matpe	0.8	0.5	−37.5	1.8	1.2	−33.3
Peas	0.2	0.7	250.0	0.4	0.4	0.0
Total pulses	7.9	5.6	−29.1	15.8	10.4	−34.2

Source: NSS data, various rounds.

The regional differences in preference for different pulses are evident in Table 2A.3 in the chapter appendix. Although pigeon pea is the most preferred pulse in the states of Andhra Pradesh, Gujarat, Karnataka, Madhya Pradesh, Maharashtra, and Chhattisgarh, chickpea is the most favored pulse in the states of Delhi, Haryana, Himachal Pradesh, Jammu and Kashmir, Kerala, Punjab, and Rajasthan. Lentil is the preferred pulse throughout the eastern region of the country, with most of the eastern and northeastern states having a preference for this pulse. Based on the data provided in Table 2A.3, a ranking of the top two most preferred pulses in different states of India reveals that pigeon pea and chickpea are the two most preferred pulses in all states, except in some eastern and northeastern states, where the most preferred pulse is lentil, followed by green gram.

Trends in Overall Protein Intake

As discussed earlier, pulses are a widely accepted ingredient in the Indian diet for the richness of their protein content. As a consequence of the changing patterns of food consumption, however, the typical protein intake in India has changed as well. Total protein intake from all sources has averaged between 60 grams and 70 grams per capita per day, and the long-term trend has been a declining one. Total protein intake averaged 61.9 grams (per person per day) in 1988–1989, then 61.6 grams in 1999–2000, 63.8 grams in 2009–2010, and most recently (per the latest round of NSS data in 2011–2012) 57.0 grams in the rural sector and 55.6 grams in the urban sector (Table 2.5). Overall in rural India as a whole, protein intake per person per day has definitely declined since 1993–1994. However, at the all-India level, the decline shows signs of flattening out, being only 0.5 gram per person per day less in 2011–2012 than it was in 2004–2005.

As expected, the protein intake of poor and middle-income households has long been lower than that of rich households. In 2009–2010, per capita daily protein consumption was 47.8 grams among poor households, 58.9 grams among middle-income households, and 84.7 grams among rich households. Such differentials arise due to disparities in the levels, type, and quality of food consumed by the households of different income groups. Based on the 68th round of the National Sample Survey (NSS), average daily protein intake rises steadily with the monthly per capita consumption expenditure (MPCE). In rural India, as of 2009–2010, it was 43 grams (per capita per day) for the bottom 5 percent of the population as ranked by MPCE and more than double that amount, 91 grams (per capita per day), for the top 5 percent. In urban India, it was 44 grams (per capita per day) for the bottom 5 percent and 87 grams (per capita per day) for the top 5 percent. The two decades between 1988–1989 and 2011–2012 saw a decline in protein intake that affected urban and rural households alike. The poor and middle-income households have not recovered their 1988 levels of protein. Between 1988–1989 and 2011–2012, protein intake fell among urban households by more than 5 percentage points among the poor, by 8.2 percentage points among the middle-income households, and by 10.6 percentage points among the rich.

Wide Variations by Region

Based on the most recent NSS data (68th round, 2011–2012), these declines in protein intake also show significant variation across space. For example,

TABLE 2.5 Trends in protein intake by households of different income groups in India, 1988–2011 (in grams per capita per day)

Income group of households	1988	1999	2011	Change, % (1988–2011)
Rural India				
Poor	50.6	44.7	44.6	−11.8
Middle income	64.0	56.0	53.1	−17.0
Rich	84.2	72.2	65.0	−20.8
All	66.4	61.1	57.0	−14.2
Urban India				
Poor	48.6	44.8	43.4	10.7
Middle income	57.9	54.7	49.6	−14.3
Rich	73.0	73.8	62.2	−14.8
All	60.3	62.7	55.6	−7.8
All India				
Poor	49.9	44.7	44.1	−11.6
Middle income	62.2	56.3	52.2	−16.1
Rich	80.3	73.5	63.8	−20.5
All	64.4	61.6	56.5	−12.3

Source: Authors' calculation based on NSS data.

Note: Poor = below poverty line; middle income = between 100 percent and 150 percent of poverty line; rich = above 150 percent of poverty line. The All India figures from NSS report 560 for the latest round 68 are used and cover the major states only.

in the rural sector, per capita intake per day varied from about 52 grams (Chhattisgarh) to about 73 grams (Haryana), and in urban areas, it varied from 55 grams (Assam) to about 69 grams (Haryana). In some of the poorer states, protein intake was markedly lower in the rural sector than in the urban sector; examples are Jharkhand (rural, 54.7 grams; urban, 60.3 grams) and Chhattisgarh (rural, 51.7 grams; urban, 55.8 grams). In the states with the highest levels of protein intake, specifically Haryana, Rajasthan, and Punjab, it was the rural population that had higher protein intake, about 4 grams to 5 grams higher than the urban population (NSSO 2014a and NSSO 2014b).

Given the country's widespread vegetarianism, apart from pulses, the two chief sources of protein for most people are cereals and dairy (milk and milk products). The relative share that these two food groups contribute to protein consumption varies somewhat by rural-urban location and varies considerably by region. The most recent NSS data (2011–2012) show that cereals make up 58 percent of rural and 49 percent of urban Indians' protein intake. Dairy products make up 10 percent of protein intake in rural India and 12 percent in urban India, but this varies significantly by state. The dairy fraction of protein is highest in the northern states of Haryana (rural, 27 percent; urban, 22 percent) and Punjab (both rural and urban, 23 percent), and ranges between 14 percent and 18 percent in Rajasthan and Gujarat. Among the 17 major states, those 4 along with Uttar Pradesh (rural, 11 percent; urban, 13 percent) were the only 5 states where the contribution of milk and milk products was higher than the national average.

The contribution of different food groups to protein intake of rural and urban India over time is presented in Figure 2.1. As mentioned, cereals are the major contributor, followed by pulses. Cereals make up more than half the share in the total per capita protein intake, although that share has been consistently declining over the years. Cereals are only a moderate source of protein, because they contain about 10 percent protein by weight. Rice contains less protein (7 percent) than wheat (approximately 12 percent) and other cereals. Over time, the contribution of cereals to protein intake has fallen. Over the past two decades, based on NSS data up to 2011–2012, the contribution of cereals to total protein intake by rural households was nearly 70 percent in 1988–1989, and this fell to 64.5 percent in 1999–2000 and fell further to 58 percent by 2011–2012. A similar fall occurred among urban households, for whom the share of cereals in protein fell from 60.4 percent in 1988–1989 to 49 percent in 2011–2012 (among the major states).

Analysis by income fractile shows another clear pattern. The contribution of cereals to protein intake falls steadily as MPCE rises: in rural India it falls from 72 percent of protein intake for the bottom 5 percent of the population to 42 percent for the top 5 percent, and in urban India it falls from 68 percent (bottom) to 31 percent (top) (NSSO 2014b). The contribution of milk and milk products follows the opposite pattern: in the rural sector it rises from 3 percent of protein for the bottom income group to 16 percent for the top group, and in the urban sector it rises from 4 percent (bottom) to 17 percent (top).

Excluding dairy, animal source foods (ASF) play a minor role in most of the country's protein diet. As of 2011–2012, the share of meat-fish-eggs (MFE) in protein intake was only 7 percent in rural India and 9 percent in urban India. This too varies greatly by region: the share was 26 percent in both rural and urban Kerala, and it was 10 percent or more in five other major states: Andhra Pradesh, Assam, Karnataka, Tamil Nadu, and West Bengal.

FIGURE 2.1 Trends in contribution of different food groups to protein intake of rural and urban households in India, 1993–1994 to 2011–2012 (percentage of share in total protein intake)

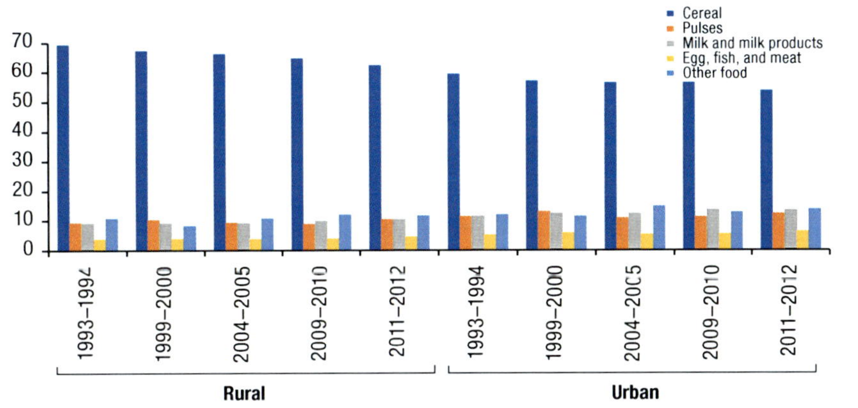

Source: NSSO (2014b).

Trends in Protein Intake from Pulses

As Figure 2.1 shows, pulses on average have been the second highest contributor to Indians' dietary protein, contributing 10 percent of protein intake in 2011–2012. Note that some of the processed items (part of the "other food" category), such as the popular chickpea flour, also include pulses. Consumers are increasingly substituting the traditional sources of proteins with alternative high-value food sources. The consumption of processed foods—such as biscuits, salted refreshments, prepared sweets, cooked meals, cake, and pastry—now contributes substantially to protein intake.

Pulses' fractional contribution to protein varied slightly between urban and rural zones, accounting for 10 percent of protein in rural India and 11 percent in urban India. Within the rural sector, this contribution ranged between 9 percent and 12 percent in 12 of the major states, while in the urban sector it ranged between 10 percent and 13 percent in 13 of the major states. Pulses' contribution has also varied over time. Starting from a level of 11.5 grams (per capita per day) in 1988, it increased slightly to 12.9 grams by 1999–2000 and then dropped to 7.4 grams by 2009–2010 and marginally recovered to 7.7 grams in 2011–2012. This depicts a notable fall of 35.7 percent over the 1988–2009 period (NSSO 2014b).

The disparities in protein intake from pulses across households of different income classes are most apparent for the years 1988–1989 and 1999–2000 in both rural and urban India. As illustrated in Table 2.6, the absolute values of protein from pulses in 2011–2012 ranged from 7 grams to 7.7 grams (per capita per day) across different income groups in rural India, and from 7.7 grams to 8.1 grams across different income groups in urban India, for a countrywide total of 7.3 grams to 7.7 grams. Hence, according to the latest NSS data, the differentials across income groups have further narrowed. Yet, overall, the 2011–2012 data indicate that the bottom 50 percent of the Indian population suffered from protein deficiency, as they consumed on average less than the recommended 60 grams per capita per day (Table T11 of NSSO 2014b).

State by State Trends in Protein Intake from Pulses

Protein intake from pulses declined across both rural and urban households during the two-decade period of 1988–2009 in every state except Kerala. Among rural households, at the beginning of the period, protein from pulses had ranged in the different states from 4 grams to 10 grams (per capita per day), and by the end of the period it had declined to 3 grams to 9 grams. Among urban households, protein from pulses declined during the period from 4 grams to 12 grams to 3 grams to 9 grams (Table 2A.4 in the chapter appendix).

Most of the eastern states experienced a decline of more than 50 percent in the contribution of pulses to dietary protein. Even some large states (such as Bihar, Madhya Pradesh, and Rajasthan) experienced a decline of around 40 percent. The differentials across rural-urban and poor-rich households also varied across states (Table 2A.5). In general, most states showed a declining trend in protein intake from pulses over the period, in both absolute terms and as a share of total protein intake. In the latest NSSO round (2011–2012) there has been a slight change with the contribution of pulses to protein reaching 10 percent at all-India level.

Table 2.7 shows that the contribution of pulses to protein intake likely depends on several factors. In the richer states, such as Andhra Pradesh and Karnataka, cereals play a substantially smaller role in delivering protein, while in the poorer states, like Chhattisgarh and Madhya Pradesh, both cereals and pulses generally make up a higher contribution to protein. In coastal states, fish make a higher contribution to protein consumption, while at the same time widespread vegetarianism in states such as Gujarat results in a comparatively high share of pulses in protein. And in northern states, where the dairy

TABLE 2.6 Trends in protein intake from pulses across households of different income groups in India, 1988–2011 (in grams per capita per day)

Income group of households	1988	1999	2011	Change (%) (1988–2011)
Rural India				
Poor	9.3	9.0	7.0	–24.7
Middle income	10.5	11.2	7.3	–30.4
Rich	12.0	13.0	7.7	–35.8
All	10.8	11.8	7.4	–31.5
Urban India				
Poor	11.2	11.0	7.7	–31.3
Middle income	12.8	13.9	7.9	–38.3
Rich	14.1	16.1	8.1	–42.5
All	12.9	14.8	8.0	–38.0
All India				
Poor	9.9	9.8	7.3	–26.3
Middle income	11.1	12.0	7.6	–31.5
Rich	12.7	14.3	7.9	–37.8
All	11.5	12.9	7.7	–33.0

Source: Authors' calculations based on NSS data.

Note: The numbers in this table are slightly higher than those implied by Tables T10-T12 of NSSO 2014b because some outliers were dropped, and some cleaning of the NSS data was done (for example, deleting households where the consumption of cereals was 0 and calorie consumption was greater than 5500 kcal per day per capita). Poor = below poverty line; middle income = between 100 percent and 150 percent of poverty line; rich = above 150 percent of poverty line.

TABLE 2.7 Percentage of contribution of different food items to protein intake across major states, rural and urban areas (%)

State	Cereals	Pulses	Milk and milk products	Meat-fish-eggs	Other food
Rural areas					
Andhra Pradesh	51	11	8	12	18
Assam	59	10	4	14	14
Bihar	62	9	9	6	15
Chhattisgarh	63	12	2	6	18
Gujarat	56	12	14	3	15
Haryana	51	8	27	2	13

(continued)

TABLE 2.7 *(continued)*

State	Cereals	Pulses	Milk and milk products	Meat-fish-eggs	Other food
Jharkhand	62	8	5	6	18
Karnataka	49	12	8	10	21
Kerala	37	9	7	26	21
Madhya Pradesh	66	10	8	2	14
Maharashtra	55	12	7	6	21
Odisha	65	9	3	7	17
Punjab	51	10	23	1	15
Rajasthan	64	6	17	1	12
Tamil Nadu	47	14	10	12	18
Uttar Pradesh	64	10	11	3	13
West Bengal	58	7	4	14	18
All	58	10	10	7	16
Urban areas					
Andhra Pradesh	46	12	11	12	19
Assam	52	11	6	16	15
Bihar	61	10	9	6	15
Chhattisgarh	56	13	6	6	19
Gujarat	50	13	17	3	18
Haryana	47	10	22	2	20
Jharkhand	56	11	9	8	17
Karnataka	42	13	11	11	23
Kerala	35	9	8	26	21
Madhya Pradesh	60	11	11	3	15
Maharashtra	46	12	11	9	22
Odisha	56	10	7	8	19
Punjab	50	11	23	2	15
Rajasthan	59	7	18	2	15
Tamil Nadu	41	14	12	12	21
Uttar Pradesh	57	11	13	5	15
West Bengal	47	8	7	19	19
All	49	11	12	9	18

Source: NSSO (2014a) and NSSO (2014b).

Note: "Major states" = those with populations greater than 20 million.

sector is more developed (such as Gujarat, Haryana, and Punjab), the dairy contribution to protein is substantially higher.

Consumer Subsidies: Not Likely to Mitigate Low Intake of Protein from Pulses

Pulses are characterized by the unique situation in which the per capita consumption of pulses has declined continuously while aggregate demand has simultaneously increased due to rapid population growth. Analysis of the NSS data shows that over time there has been a reduction in per capita consumption of pulses across all household categories (by wealth) and across all regions of the country, part of which can be explained by high prices of pulses. Going forward, the price management in pulses is therefore likely to be quite important. In this regard, could the public distribution of pulses at subsidized prices help the consumption of pulses? The PDS (public distribution system) is jointly operated by India's central and state governments. That joint operation provides some flexibility to the state governments regarding within-state allocation as well as choosing the product mix. Traditionally, India's PDS provides subsidized rice and wheat (along with kerosene and sugar in some places) through a nationwide network of fair price shops.

The recent National Food Security Act of 2013 made 25 kilograms of grains available to two-thirds of households in India at highly subsidized prices of 2 rupees per kilogram for wheat and 3 rupees per kilogram for rice. Furthermore, a system of highly subsidized grains called the Antyodaya Anna Yojana (AAY) was introduced for the poorest households.

When a commodity is included in the PDS, different aspects related to implementation become important. The inclusion of pulses in the PDS does not have such detailed allocation rules. Specifically, regarding pulses, they form a group with different types and varieties. Moreover, there is heterogeneity in preferences across Indian states. Only specific types of pulses are consumed in each region, with little substitution among them. Some states that have included pulses in the PDS have tended to keep the state's most preferred pulses in the subsidy plan. Table 2.8 shows significant price increases (measured in terms of unit values from NSS data) in pulses over time in the open market. The price increases have clearly been significant, and one would expect a consumption subsidy to be important with such price dynamics.

The inclusion of pulses has to specify some allocation and assignment rules as with rice and wheat. The rule can be individual based (quantity per

TABLE 2.8 Pulse prices in states including it in the PDS and those not including it

Pulse variety	2004–2005 constant prices (rupees per kilogram)		Increase in prices (%)
	2004–2005	2009–2010	
Pigeon pea	30.11	46.14	53.25
Chickpea	24.93	25.19	1.03
Green gram	28.54	43.79	53.42
Lentil	28.78	40.75	41.60
Black matpe	26.87	37.79	40.63
Peas	20.22	20.69	2.32
Other pulses	26.13	32.01	22.51
All	27.63	36.58	32.40

Source: NSSO consumption expenditure data corresponds to 2004–2005 and 2009–2010.

member of household), or it can be specified at the household level with its affiliated price. As is evident in Table 2.9, states have different arrangements with regard to the inclusion of pulses in the PDS. While Andhra Pradesh and Tamil Nadu exclusively include pigeon pea in their PDS, Himachal Pradesh and Punjab have introduced a mix of pulses in the subsidy program.

With the inclusion of pulses in the PDS in some states as outlined here, Chakrabarti, Kishore, and Roy (2016) show only limited impact of the inclusion of pulses in the PDS on household consumption as increased PDS uptake is matched by shrinkage in market purchases. Note that given the deficit in pulses, only limited amounts (that is, 0.5 kilogram to 2 kilograms) of pulses have been provided by the selected states through the PDS. Overall, consumer price subsidy in pulses introduced in different states did not result in significantly improved nutrition in terms of household protein intake. This result is despite the fact that the households in their sample are poor and many are vegetarians and pulses have been subject to significant price rises—a situation often blamed for the falling or stagnant consumption of pulses. India remains a country where malnourishment is widespread, including protein deficiency, for which consumption of pulses could be a mitigating factor. This state of affairs was the principal motivation for the inclusion of pulses in India's PDS in the first place.

Notwithstanding the importance of the issue, the inclusion of pulses in the PDS might not be the way forward. With a subsidy, channels such as

TABLE 2.9 Pulse subsidization scheme in different states

State	Pulse subsidized	Details	Year of introduction of pulses in the public distribution system
Andhra Pradesh	Pigeon pea	1 kilogram at 50 rupees per kilogram per household per month	2008
Himachal Pradesh	Moong whole	1 kilogram per ration card having 5 and above family members per month at 49.99 rupees per kilogram	2007
Himachal Pradesh	Urd dal	1 kilogram per ration card per month to all ration card holders at 34.99 rupees per kilogram	2007
Himachal Pradesh	Chana dal	1 kilogram per ration card having 3 and above family members per month at 25 rupees per kilogram	2007
Punjab	Various pulses	0.5 kilogram per member to a maximum of 2.5 kilograms per family per month at 20 rupees per kilogram	2007
Tamil Nadu (two pulses, arhar dal and urd dal)	Pigeon pea	1 kilogram at 30 rupees per kilogram per household per month	2007
Tamil Nadu	Urd dal	1 kilogram at 30 rupees per kilogram per household per month	2007

Source: Chakrabarti, Kishore, and Roy (2016).

substitution and wealth effects come into play that could lead to end results different from what were primarily expected. Whereas the subsidies do appear to have affected pulse consumption in a statistically significant way, the size of the effect is not large enough to make much difference nutritionally. The increase in consumption is at best less than 30 percent of the incrementally subsidized amount. Essentially it seems that subsidies have induced substitution away from pulses to other food and even nonfood items that need not address the protein needs of the population. The estimates show that application of a subsidy to a commodity, the preference for which is declining, is unlikely to affect several of the outcomes that matter.

If the counterfactual would have been greater reduction in consumption of pulses, then policy makers should be satisfied, but improved nutrition seems farfetched in the case of a commodity that is losing favor with consumers. The results in Chakrabarti, Kishore, and Roy (2016) showing little nutritional effects from the consumption subsidy have been established in studies other than that of Jensen and Miller (2011) for different nutrients in India (see Kochar 2005; Tarozzi 2005; Behrman and Deolalikar 1988).

Demand Projections for Pulses to 2020 and 2030

In the preceding pages, we have laid out a picture of the changing per capita demand for pulses over time, which shows a general decline in per capita consumption. Those numbers, however, do not reveal a secondary pattern: Despite declining per capita consumption, the total demand for pulses and processed foods derived from them has actually been increasing. This increase stems from many factors, including an increase in population, rising household income, rising numbers of two-earner couples, product diversification, and greater urbanization and the lifestyle changes associated with it, such as changing tastes and preferences. Taking these factors into consideration, next we look to the future demand for pulses in India to the years 2020 and 2030.

Methodology

The demand for pulses comprises both direct and indirect demand. The direct demand consists of consumption at home and outside the home. The indirect demand includes pulses' use as seed and animal feed, in industry, as well as loss of pulses in wastage and spoilage. Here, we estimate the future demand for pulses at the disaggregated level based on projections for consumers' income levels, geographic location (rural and urban households), and location in terms of states–union territories (UTs). To capture the effects of different determinants of demand, we classified the rural-urban households of 35 Indian states-UTs into the three income groups described above. The baseline consumption of pulses, their demand elasticity, the rise in consumer income, and population growth are the four most important factors behind our demand projections (see Mittal 2010; Kumar, Shinoj, and Raju 2011). The growth rates in per capita income were obtained by subtracting the population growth rate from gross domestic product (GDP) growth and were used in predicting per capita consumption. The estimated per capita consumption was multiplied by the projected population and aggregated by the states-UTs, income groups, and rural-urban households to arrive at household demand at the national level.

Measuring pulse demand at home. The per capita pulse demand at the household level can be predicted by equation (1):

$$d_{ijkt} = d_{ijkt-1}\,[1 + y_{ijt}\,.\,e_{ijk}(1 - s)] \qquad (1)$$

The total household demand can be obtained by multiplying the per capita pulse demand at the household level by the population (N_{ijkt}), that is,

$$D_{ijkt} = d_{ijkt}\,.\,N_{ijkt} \qquad (2)$$

The aggregate household demand at the national level can be obtained by summing income groups, rural-urban households, and states-UTs, as in equation (3):

$$D_t = \Sigma i\,\Sigma j\,\Sigma k\,D_{ijkt} \qquad (3)$$

where,

d = per capita consumption of a pulse,

e = expenditure elasticity of the pulse,

s = saving rate assumed at 36 percent, as estimated by India's Central Statistical Organization (CSO),

N = population,

y = per capita GDP growth,

D = total household demand for pulses,

i = demographic status (1 for rural and 2 for urban households),

j = state or UT (ranging from 1 to 35),

k = income group of households (ranging from 1 to 3),

t = year, and

t_0 = base year (2004–2005).

The expenditure elasticities were estimated at the regional levels using the Food Characteristic Demand System, following Bouis and Haddad (1992). These regional expenditure elasticities were superimposed on the corresponding state-UT. The aggregate household human demand for the jth state-UT in the year t was computed by summing i rural-urban households and k income groups. The summation over the states-UTs provided the household demand at the national level for a pulse in the year t.

Measuring outside-of-home pulse demand. The pulse consumption of a household away from home was estimated based on the FAO Food Balance Sheet. In this approach the total pulse consumption (C) was obtained by equation (4):

$$C = (Q + M + S) - E - (\text{seed} + \text{feed} + \text{wastages} + \text{industrial uses}), \quad (4)$$

where,

Q = total production of a pulse,

M = imports of that pulse,

E = exports of that pulse, and

S = stock changes of that pulse.

The value of C included pulse consumption at home (*H*) and outside of home (*OH*). The NSS survey data on household consumption were used to estimate *H* and the pulse consumption outside of home was obtained by subtracting *H* from *C*. The baseline per capita pulse consumption outside home (O_{ijkt0}) for the subgroup *i* rural-urban households in the *j*th state of *k* income group in the base year was computed as in equation (5):

$$O_{ijkt0} = d_{ijkt0} \cdot (OH/H). \quad (5)$$

The per capita outside-of-home demand for pulses was predicted by equation (6):

$$O_{ijkt} = O_{ijk\,t-1}\,[1 + y_{ij} \cdot f_{eijk}(1-s)], \quad (6)$$

where f_e = expenditure demand elasticity for pulses consumed away from home, computed as the weighted average of demand elasticities of all the food commodities (the weights were the share of each food commodity in total food expenditure).

The total outside-of-home demand for pulses was obtained by equation (7):

$$OH_{ijkt} = O_{ijkt} \cdot N_{ijkt.} \quad (7)$$

The outside-of-home demand for pulses at the country level was computed by the summation of all the disaggregated demands, as in equation (8):

$$OH_t = \Sigma_i\, \Sigma_j\, \Sigma_k\, OH_{ijkt}, \quad (8)$$

where

Oh = per capita outside-of-home consumption of pulses,

OH = total home-away consumption of pulses.

Measuring total pulse demand. The summation of pulse demand at home and outside of home provided the total pulse demand, namely:

$$PD_t = D_t + OH_{t,} \quad (9)$$

where PD is the total demand for pulses.

Population projections. For population projections, the data provided by India's Registrar General of Census were used. These data provide the numbers for the rural and urban population by state and UT. The rural and urban populations were further split into three income groups for each state-UT by using the weights derived from the sample households of the 61st NSS round.

Income growth. For income growth, the data provided by India's Central Statistical Organization were used. These data provided the gross domestic product at factor cost for both the agricultural and allied-activities sector and the national economy at 1999–2000 prices. From these data series, the five-year moving average for growth in agricultural, nonagricultural, and total economic activities was computed up to 2009. The economy's slowdown in 2008 and the start of its pickup in 2009 were assumed with the 25 percent growth recovery in 2010 and 2011, and economic growth was assumed to be constant through the projection year of 2030. The agricultural GDP growth rate was assumed as the income growth rate for rural households, and the nonagricultural GDP growth rate was assumed as the income growth rate for urban households.

Measuring indirect demand for pulses. The indirect demand for pulses comprises their use as seed and animal feed, their use in industry, and the share of food wastage and was estimated as follows:

Seed. The seed requirement was estimated based on projected area under pulse cultivation and the application of seed rates.

Feed. The demand for feed grains for livestock consumption was computed using their demand for livestock products in terms of livestock output units (LOUs) and the average feeding ratio—that is, the quantity of feed required per unit of livestock products. The LOU was worked out by adding the required quantities of meat and eggs and one-tenth of milk. Looking at the importance of aquaculture, one-tenth of fish production was also included in the LOU (Kumar 1998). The feed demand was estimated by multiplying the LOU by the feeding ratio (feed required per unit of LOU). The feed requirement is largely met by food grains, oilcakes, and cotton seed. Pulse grains contributed about 5.1 percent to the total feed demand.

Industrial uses. An industrial-use allowance of 5 million tons of food grains, as provided by the National Commission on Agriculture in 2000, was used. In the total industrial use of food grains, pulses accounted for about 7 percent. The industrial use for pulses was projected assuming an annual growth rate of 3 percent.

Wastage. India's Directorate of Economics and Statistics, Ministry of Agriculture, assumes that 2.2 percent of pulse production is lost as wastage. This wastage allowance includes the pulse grains not fit for human consumption and those used as feed. To address the problem of double accounting, only half of this allowance was accounted as feed in this study.

Estimation of Indirect Consumption Demand for Pulses

The FAO Food Balance Sheet for commodities estimates the share of indirect demand in total supply (production + import + change in stocks) as 16.4 percent for pulses. A study has estimated the share of indirect demand for pulses in total demand as 15.2 percent to 15.5 percent (Kumar 1998). The Directorate of Economics and Statistics considers the share of seed, feed, and wastage in pulse production to be 12.5 percent. In the present study, the share of indirect demand (seed, feed, wastage, and industrial use) in total pulse demand was estimated to be 18.9 percent. The use of pulses for human consumption was estimated as 81.1 percent of the total demand (62.9 percent at home and 18.2 percent outside of home). Table 2.10 provides details.

Different Income Growth Scenarios

The annual per capita consumption and total demand for pulses have been projected in the study under the following three scenarios:

S1 (constant GDP growth scenario) = Existing GDP growth and constant expenditure elasticity in the projected period

S2 (low GDP growth scenario) = 25 percent lower GDP growth and constant expenditure elasticity in the projected period, and

S3 (high GDP growth scenario) = 25 percent higher GDP growth and constant expenditure elasticity in the projected period.

TABLE 2.10 Sources of total demand for pulses in India

Demand group	Demand source	Pulses share (%)
Household demand	At home	62.9
	Home-away	18.2
	Subtotal	81.1
Indirect demand	Seed	6.7
	Feed	8.7
	Wastages	1.1
	Industrial use	2.5
	Subtotal	19.0
Total demand		100.1

Source: Authors' calculations based on FAO Food Balance Sheet. from FAOstat.

Demand Elasticity for Pulses

In this study, the demand elasticities for pulses at the state-UT level, for rural-urban households, and for different income groups were all computed using the Food Characteristics Demand System (FCDS) model (Bouis and Haddad 1992) based on NSS data pertaining to the year 2009. The national-level estimates of income and own-price elasticities were computed as the weighted averages of the disaggregated elasticities and are presented in Table 2.11. As Table 2.11 shows, the demand elasticities varied widely across demographic areas (rural-urban) and income groups. The magnitude of elasticity declined with the rise in income across all income groups, and the estimates are slightly higher for rural than for urban households. The own-price elasticities had the expected negative sign.

Pulse Demand for Human Consumption

As detailed in the methodology, the pulse demand for human consumption was computed by multiplying the projected per capita consumption by the projected population at two stages: (1) at home and (2) outside home. These projections were computed under the three income-growth scenarios (described above) at 10-year intervals, from 2010 to 2030, and are presented in Table 2.12.

Pulse consumption at home in the base year (2010) has been estimated at 11.33 million tons. This demand is projected to rise to 13.48 million tons to 14.07 million tons by 2020 and further to 15.82 million tons to

TABLE 2.11 Demand elasticity for pulses in India, 2009

Demographic area	Income group of households: Poor	Middle income	Rich	All households
Expenditure elasticity				
Rural India	0.499	0.285	0.111	0.248
Urban India	0.501	0.260	0.090	0.176
All India	0.500	0.274	0.098	0.206
Own-price elasticity				
Rural India	−0.686	−0.507	−0.300	−0.448
Urban India	−0.723	−0.557	−0.381	−0.462
All India	−0.699	−0.530	−0.349	−0.456

Source: Authors' estimations.

Note: Poor = below poverty line; middle income = between 100 percent and 150 percent of poverty line; rich = above 150 percent of poverty line.

TABLE 2.12 Demand projections for pulses in India under different income growth scenarios, 2010–2030 (in millions of metric tons)

Year	Current GDP growth (S1)	Low GDP growth (S2)	High GDP growth (S3)
Demand at home			
2010	11.33	11.33	11.33
2020	13.77	13.48	14.07
2030	16.64	15.82	17.54
Demand outside of home			
2010	3.28	3.28	3.28
2020	4.05	3.95	4.15
2030	4.98	4.69	5.30
Total direct demand (at home + outside home)			
2010	14.61	14.61	14.61
2020	17.81	17.42	18.22
2030	21.62	20.51	22.84
Indirect demand (seed, feed, wastage, and other uses)			
2010	3.41	3.41	3.41
2020	4.06	3.98	4.14
2030	4.95	4.71	5.23
Total domestic demand for pulses			
2010	18.02	18.02	18.02
2020	21.87	21.40	22.36
2030	26.58	25.22	28.07

Source: Authors' estimations.

17.54 million tons by 2030. The outside-home consumption is projected to rise to 3.95 million tons to 4.15 million tons by 2020 and to 4.69 million tons to 5.30 million tons by 2030. The total human demand for pulses at home and outside home has been projected to reach 17.42 million tons to 18.22 million tons by 2020 and 20.51 million tons to 22.84 million tons by 2030.

Indirect Demand for Pulses

Next, we project the demand for pulses as seed, animal feed, and in industrial uses and loss as wastage for the same period: 2010–2030. For 2010 we estimate the annual requirement of pulses as seed at 1.2 million tons. We project the demand for pulses as animal feed at 1.94 million tons to 2.09 million tons by 2020 and 2.41 million tons to 2.91 million tons by 2030. These projections are based on a feeding ratio of 1.5, starting with an estimate of the 2010 demand for all animal feed (from all sources, including pulses) as 30.7 million tons, which we project to grow to 38 million tons to 41 million tons by 2020 and to 47 million tons to 57 million tons by 2030. We estimate pulses to account for about 5.1 percent of those total animal feed requirements. A higher demand for pulses is expected for feed use because of the shift in dietary patterns toward consumption of more livestock products as incomes rise and urbanization increases.

We project the demand for pulses in industrial uses to grow to 0.61 million tons by 2020 and to 0.82 million tons by 2030, based on an annual growth rate of 3.1 percent. Concerning wastage, we estimate the loss of pulses to be 0.20 million tons in 2010 and project it to rise to 0.24 million tons by 2020 and to 0.29 million tons by 2030. Consequently, our overall projected demand for pulses as seed, animal feed, and industrial uses combined, along with loss as wastage, is computed to rise to 3.98 million tons to 4.14 million tons by 2020 and to reach 4.71 million tons to 5.23 million tons by 2030.

Total Demand Projections for Pulses to 2030

Adding our projections for direct human consumption of pulses to our projections for indirect demand, we arrive at a projection of total demand for pulses of 22 million tons by 2020 and 25 million to 28 million tons by 2030 (Table 2.12). This projected demand implies an annual growth rate of 1.73 percent to 2.18 percent during the 2010–2020 period and of 1.66 percent to 2.30 percent during 2020–2030. Note that these demand projections entail a constant price forecast and deficit (demand supply gap) as estimated, and they could be altered by higher prices that could suppress demand while at the same time possibly augment supply.[3] Going forward, other factors, such as increased imports in combination with changes in production, could also alter the net availability of pulses.

Conclusion

In the context of the diversification of Indian consumers' diets, this chapter has focused on the long-term changes in the patterns of their consumption of pulses and has underscored the fact that households' consumption of cereals and pulses generally declined during the past two decades. Between 1988 and 2009–2010, the decline in pulse consumption has been appreciable at 26.7 percent, which represents a 1.47 percent decline annually. Data from 2011–2012 show a modest turnaround after passing through a period of drought, but this uptick has in no way been large enough to reach the consumption levels of two decades ago.

Across the different pulses, chickpea and pigeon pea, which constitute nearly half of all pulse consumption in India, have experienced a considerable drop in consumption levels. The consumption of some other major pulses, such as green gram, lentil, and black matpe, has also declined in recent years. However, some less expensive pulses, such as peas and lentils, have gained in their consumption shares over the years, which implies some substitution among pulses. On further exploration, the level of substitution between pulses and high-value food commodities has been found comparable across rural and urban households, although the former has shown a slightly higher tendency to move away from pulses.

Pulses' contribution as a major source of proteins has also changed across regions and over time. In some places, such high-value commodities as milk, meat, eggs, and processed products have emerged as substitutes for pulses in supplying dietary protein. These trends are observed irrespective of household demographics or income, although intergroup disparities are evident. Nevertheless, making up more than 10 percent of the protein diet in the country overall, and in some states as much as 14 percent, pulses remain a sizable contributor to the protein intake of Indian households. Although per capita consumption of pulses has declined over time, the total demand for pulses

3 References to *demand-supply gap* here and elsewhere in the book pertain to the gap between domestic demand and domestic production.

has continuously grown in India, driven by the rising population, growing economy, and expanding urbanization. This study has projected the future demand of pulses by (and the projections under the baseline have been undertaken by) considering household consumption demand, outside-of-home consumption, and indirect demand such as seed, animal feed, and industrial use.

The projections under the baseline scenario indicate that the domestic demand for pulses will grow from the present level of 18.00 million tons to 21.9 million tons by 2020 and to 26.6 million tons by 2030. Under an alternative scenario of low GDP growth (25 percent lower than baseline), our projections are 21.4 million tons and 25.2 million tons, respectively, by 2020 and 2030. Under the third scenario of high GDP growth (25 percent higher than baseline), our projection estimates demand to reach 22.3 million tons by 2020 and 28.1 million tons by 2030.[4]

To meet this projected demand, proactive steps will certainly be required to augment pulse production in India. Unless that is done, the country would have to meet the domestic demand for pulses by increasing imports. Furthermore, given the higher protein deficiency among lower income strata and the high pulse price elasticity of poor consumers, there is a rationale for policies that can contribute to lower consumer prices such as reducing pulse processing costs through the promotion of improved technology, reducing intermediation costs at various levels by facilitating direct links between farmers and processors, and reducing farmers' production costs through improved technology. These issues are discussed in subsequent chapters.

References

Alwan, A., D. R. McLean, L. M. Riley, E. T. d'Espaignet, C. D. Mathers, G. A. Stevens, and D. Bettcher. 2010. "Monitoring and Surveillance of Chronic Non-communicable Diseases: Progress and Capacity in High-burden Countries." *Lancet* 376: 1861–1868.

Behrman, J. R., and A. B. Deolalikar. 1988. "Health and Nutrition." In *Handbook of Development Economics*, edited by H. Chenery and T. N. Srinivasan, 631–711. Amsterdam: North-Holland.

Bouis, H., and L. Haddad. 1992. "Are the Estimates of Calorie Income Elasticities Too High? A Recalibration of the Plausible Range." *Journal of Development Economics* 53 (June): 115–137.

Boye J., F. Zare, and A. Pletch. 2010. "Pulse Proteins: Processing, Characterization, Functional Properties, and Applications in Food and Feed." *Food Research International* 43 (2): 414–431.

4 In fact, real prices of pulses rose significantly between 2009 (base year for the projections) and 2016, which could imply that the reported estimates are biased upward.

Chakrabarti, S., A. Kishore, and D. Roy. 2016. *Effectiveness of Food Subsidies in Raising Healthy Food Consumption: Public Distribution of Pulses in India.* IFPRI Discussion Paper 1523. Washington, DC: International Food Policy Research Institute (IFPRI).

Gopalan, C., B. V. Rama Sastri, and S. C. Balasubramanian. 1989. *Nutritive Value of Indian Foods.* Hyderabad, India: National Institute of Nutrition.

Jensen, R. T., and N. H. Miller. 2011. "Do Consumer Price Subsidies Really Improve Nutrition?" *Review of Economics and Statistics* 93 (4): 1205–1223.

Joshi, P. K., and P. P. Rao. 2011. *Food Demand and Supply Projections for India: 2010–2030.* Trade, Agricultural Policies and Structural Changes in India's Agrifood System; Implications for National and Global Markets (TAPSIM) Project KBBE212617. New Delhi: International Food Policy Research Institute (IFPRI).

———. 2017. "Global Pulses Scenario: Status and Outlook." *Annals of New York Academy of Sciences*, 1392: 6–17. doi:10.1111/nyas.13298.

Jukanti, A. K., P. M. Gaur, C. L. L. Gowda, and R. N. Chibbar. 2012. "Nutritional Quality and Health Benefits of Chickpea (*Cicer arietinum L.*): A Review." *British Journal of Nutrition* 108: S12–S26.

Khushwaha, A., P. Rajawat, H. S. Kushwah. 2002. "Nutritional Extrude of Faba Bean (*Vicia faba L.*) As a Protein Supplement in Cereals Based Diets in Rats." *Journal of Experimental Biology* 40 (1): 49–52.

Kochar, A. 2005. "Can Targeted Food Programs Improve Nutrition? An Empirical Analysis of India's Public Distribution System." *Economic Development and Cultural Change* 54 (1): 203–235.

Kumar, P. 1998. *Food Demand and Supply Projections for India.* Agricultural Economics Policy Paper 98–01. New Delhi: Indian Agricultural Research Institute.

Kumar P. A., P. Shinoj, and S. S. Raju. 2011. "Estimation of Demand Elasticities for Food Commodities in India." *Agricultural Economics Research Review* 24 (1): 1–14.

McCrory M. A., B. R. Hamaker, J. C. Lovejoy, and P. E. Eichelsdoerfer. 2010. "Pulse Consumption, Satiety, and Weight Management." *Advances in Nutrition* 1:17–30.

Mittal, S. 2010. "Application of the QUAIDS Model to the Food Sector in India." *Journal of Quantitative Economics* 8 (1): 42–54.

NSSO (National Sample Survey Office). 2014a. *Household Consumption of Various Goods and Services in India, 2011–2012.* NSSO 68th Round. Ministry of Statistics and Program Implementation, Report 558. Accessed July 2016. http://mospi.nic.in/mospi_new/upload/Report_no558_rou68_30june14.pdf.

———. 2014b. *Nutritional Intake in India, 2011–2012*. NSSO 68th Round. Ministry of Statistics and Program Implementation, Report 560. Accessed July 2015. http://mospi.nic.in/mospi_new/upload/nss_report_560_19dec14.pdf.

Patwardhan, V. N. 1962. "Pulses and Beans in Human Nutrition." *American Journal of Clinical Nutrition* 11 (July): 12–30.

Sukhatme, P. V. 1970. "Incidence of Protein Deficiency in Relation to Different Diets in India." *British Journal of Nutrition* 24 (2): 477–487.

Tarozzi, A. 2005. "The Indian Public Distribution System as Provider of Food Security: Evidence from Child Nutrition in Andhra Pradesh." *European Economic Review* 49 (5): 1305–1330.

Vijayaraghavan, P. K., and P. R. Srinivasan. 1953. "Essential Amino Acid Composition of Some Common Indian Pulses." *Journal of Nutrition* 51 (2): 261–271.

Appendix

TABLE 2A.1 Dynamics of pulses consumption across rural and urban households in different states of India, 1988–1989 to 2011–2012 (in kilograms per capita per year)

	Rural households				Urban households			
State	1988	2009	Change (%)	2011–2012	1988	2009	Change (%)	2011–2012
Andhra Pradesh	5.6	7.9	29.11	10.28	19.2	9.96	−48.4	11.4
Assam	8.8	6.4	−27.3	7.7	12.3	7.5	−39.0	9.3
Bihar	14.1	7.2	−48.9	8.9	13.4	8.0	−40.3	9.8
Chhattisgarh	12.4	8.8	−29.0	9.5	16.9	11.5	−32.0	11.5
Delhi	14.2	8.1	−43.0	12.8	16.5	8.7	−47.3	11.1
Gujarat	11.6	8.8	−24.1	10.1	14.4	10.2	−29.2	11.4
Haryana	10.6	7.9	−25.5	9.0	11.4	8.6	−24.6	10.7
Himachal Pradesh	16.6	15.5	−6.6	15.1	20.4	15.8	−22.5	17.0
Jammu and Kashmir	8.1	6.8	−16.0	7.6	7.6	6.8	−10.5	7.8
Jharkhand	11.8	6.9	−41.5	6.8	14.0	9.0	−35.7	10.0
Karnataka	12.6	8.8	−30.2	10.9	13.1	9.9	−24.4	12.2
Kerala	6.5	7.6	16.9	8.2	7.7	8.2	6.5	9.4
Madhya Pradesh	16.0	9.1	−43.1	10.2	14.8	9.7	−34.5	11.1
Maharashtra	13.7	11.0	−19.7	11.6	15.3	11.0	−28.1	12.1
Manipur	9.7	4.8	−50.5	4.8	10.7	4.2	−60.7	4.9
Meghalaya	4.4	2.9	−34.1	3.7	—	3.8	—	6.3
Mizoram	9.9	6.0	−39.4	5.4	15.1	7.1	−53.0	7.3
Nagaland	—	4.8	—	4.9	10.1	5.2	−48.5	5.0
Odisha	6.9	7.1	2.9	7.3	10.9	8.5	−22.0	8.7
Punjab	13.2	10.1	−23.5	10.7	13.2	10.7	−18.9	11.4
Rajasthan	7.6	5.2	−31.6	6.8	9.4	5.9	−37.2	7.1
Sikkim	6.9	5.7	−17.4	5.7	10.2	7.6	−25.5	5.9
Tamil Nadu	9.7	10.8	11.3	11.9	11.9	11.7	−1.7	12.8
Tripura	7.1	5.1	−28.2	4.9	8.4	6.5	−22.6	6.6
Uttar Pradesh	15.8	9.7	−38.6	10.3	13.0	9.2	−29.2	10.6
Uttarakhand	13.3	10.2	−23.3	12.2	17.1	10.5	−38.6	11.9
West Bengal	6.3	5.2	−17.5	5.8	8.6	5.8	−32.6	6.9
India	11.2	8.1	−27.7	9.3	12.4	9.0	−27.4	10.8

Source: Authors' calculations based on different NSS rounds.

Note: — = data not available.

TABLE 2A.2 Dynamics of pulses consumption across states among poor and rich households, 1998–1999 to 2011–2012 (in kilograms per capita per year)

	NSS 55th Round			NSS 61st			NSS 66th			NSS 68th		
	1998			2003			2009			2011		
State	BPL	APL	All	BPL	APL	All	BPL	APL	All	BPL	APL	All
Andhra Pradesh	5.8	10.4	9.6	6.0	9.6	8.9	5.0	9.3	8.8	6.5	11.0	10.6
Assam	4.0	8.2	6.8	4.4	8.5	7.8	3.8	7.3	6.5	5.7	9.3	8.3
Bihar	7.1	12.4	10.4	6.2	9.9	8.5	5.1	8.0	7.0	6.9	9.9	9.1
Gujarat	7.3	11.6	11.0	6.9	9.8	9.2	5.9	9.5	8.4	7.4	9.9	9.5
Haryana	5.1	11.2	10.7	3.5	6.5	6.0	4.3	7.0	6.7	5.7	8.1	7.8
Himachal Pradesh	10.3	16.0	15.6	9.8	14.4	13.9	9.6	14.8	14.6	10.9	15.4	15.1
Jammu and Kashmir	8.5	10.2	10.1	6.7	7.3	7.3	4.6	6.3	6.2	5.5	6.7	6.6
Karnataka	8.6	13.4	12.5	7.2	10.1	9.3	6.4	9.5	8.9	7.6	11.9	11.0
Kerala	3.2	7.7	7.1	3.1	7.6	6.8	3.1	7.7	7.3	4.6	9.3	8.9
Madhya Pradesh	8.0	11.8	10.5	7.0	10.3	9.1	6.2	9.0	8.3	7.7	10.6	9.8
Maharashtra	8.4	12.4	11.4	8.0	11.2	10.2	6.7	10.5	9.8	8.5	11.4	11.0
Manipur	4.3	8.9	8.3	2.9	4.8	4.8	2.9	3.9	3.6	3.6	4.6	4.3
Meghalaya	2.5	4.2	4.1	2.6	4.0	4.0	2.1	3.0	2.9	3.2	4.5	4.4
Nagaland	2.4	8.6	8.6	—	9.6	9.6	2.5	4.1	3.9	2.7	3.9	3.8
Orissa	3.6	8.8	6.7	4.0	8.5	6.6	4.0	7.6	6.8	5.1	8.9	7.9
Punjab	8.3	12.6	12.3	6.3	9.8	9.5	6.5	9.4	9.0	7.3	9.8	9.6
Rajasthan	4.8	7.6	7.2	4.6	5.3	5.1	2.7	4.6	4.3	3.9	6.0	5.7
Sikkim	3.4	6.8	6.2	3.9	6.0	5.7	3.5	5.8	5.6	4.9	6.0	5.9
Tamil Nadu	6.5	12.0	11.0	6.4	10.9	9.7	6.8	11.1	10.6	7.9	12.3	11.8
Tripura	2.5	7.4	6.9	3.5	5.9	5.3	3.3	5.5	5.3	3.6	6.0	5.6
Uttar Pradesh	8.3	13.8	12.2	7.3	10.9	9.7	6.8	9.4	8.6	8.0	10.4	9.7
West Bengal	4.0	6.9	6.2	3.4	6.0	5.4	3.1	5.5	5.1	4.0	6.4	6.0
A and N Islands	6.4	12.2	12.1	8.2	10.2	10.2	0	11.1	11.1	4.5	11.8	11.8
Arunachal Pradesh	4.4	3.2	2.6	3.6	6.4	6.2	3.3	6.3	5.8	3.7	7.3	6.4
Chandigarh	8.5	14.8	14.6	8.3	12.1	12.0	7.2	11.6	11.2	8.5	13.9	13.7
Dadra and Nagar Haveli	11.2	15.2	14.5	11.3	11.8	11.6	10.3	13.1	12.3	11.1	15.0	13.6
Delhi	7.7	12.0	11.6	6.6	9.7	9.2	5.7	7.2	7.0	7.1	10.6	10.3
Goa and Daman and Diu	6.7	11.1	10.9	4.0	5.8	5.5	5.6	10.0	9.6	7.4	9.2	9.1
Lakshdweep	4.3	8.0	7.9	4.0	8.6	8.1	7.7	7.6	7.6	4.5	9.8	9.6

(continued)

TABLE 2A.2 *(continued)*

State	NSS 55th Round 1998 BPL	APL	All	NSS 61st 2003 BPL	APL	All	NSS 66th 2009 BPL	APL	All	NSS 68th 2011 BPL	APL	All
Mizoram	4.1	9.4	9.4	1.5	7.3	7.2	3.1	6.9	6.4	3.9	7.7	7.2
Pondicherry	6.1	11.9	10.9	6.0	11.1	9.6	4.3	11.8	11.6	7.5	13.1	12.2
Chhattisgarh	4.9	10.1	8.0	4.7	8.8	7.1	4.7	8.2	7.1	5.4	8.8	7.7
Jharkhand	6.2	10.9	9.1	6.2	12.3	10.0	4.9	9.6	8.5	7.0	11.4	10.2
Uttarakhand	8.9	13.1	12.4	9.5	12.8	11.5	6.7	9.9	9.4	9.6	12.2	11.9
India	6.5	12.3	11.0	6.1	9.0	8.3	5.2	8.3	7.7	6.4	9.5	8.9

Source: Authors' calculations based on different NSS rounds.

Note: — = data not available. BPL= below poverty line; APL = above poverty line.

TABLE 2A.3 Trends in consumption shares of major pulses in different states of India, 1988–2009 (percentage of share of total pulses consumption)

State	Major pulses	1988 (%)	1999 (%)	2009 (%)	Change (%) (1988–2009)
Arunachal Pradesh	Lentil	42.2	33.7	44.5	2.3
	Green gram	11.4	7.9	11.5	0.1
	Chickpea	10.6	5.1	13.6	3.0
	Pigeon pea	10.1	4.4	8.6	−1.5
Andhra Pradesh	Pigeon pea	46.7	53.4	54.2	7.5
	Black matpe	13.8	15.7	17.0	3.2
	Green gram	20.0	15.6	11.9	−8.1
	Chickpea	11.4	11.1	10.9	−0.5
Assam	Lentil	57.0	67.6	56.5	−0.5
	Green gram	14.3	12.9	10.7	−3.6
	Chickpea	7.2	4.9	11.0	3.8
Bihar	Lentil	34.5	41.7	31.1	−3.4
	Chickpea	21.1	23.7	33.7	12.6
	Pigeon pea	14.6	14.3	11.5	−3.1
	Green gram	6.6	9.9	10.0	3.4
Chhattisgarh	Pigeon pea	24.3	40.4	48.3	24.0
	Chickpea	17.7	6.8	18.2	0.5
	Black matpe	18.4	16.6	14.1	−4.3

State	Major pulses	1988 (%)	1999 (%)	2009 (%)	Change (%) (1988–2009)
Delhi	Chickpea	29.4	36.9	36.8	7.4
	Pigeon pea	16.3	14.5	13.6	–2.7
	Green gram	15.2	16.5	15.6	0.4
	Lentil	13.9	14.7	15.1	1.2
	Black matpe	13.0	9.3	6.5	–6.5
Gujarat	Pigeon pea	36.7	41.2	41.0	4.3
	Green gram	27.3	28.6	22.3	–5.0
	Chickpea	19.4	19.2	23.8	4.4
Haryana	Chickpea	50.9	44.4	50.9	0.0
	Green gram	18.5	19.5	17.7	–0.8
	Lentil	13.6	15.8	13.3	–0.3
	Black matpe	10.5	10.9	8.3	–2.2
Himachal Pradesh	Chickpea	36.0	36.7	44.7	8.7
	Black matpe	23.4	20.5	18.2	–5.2
	Lentil	13.9	11.6	7.0	–6.9
	Green gram	8.0	8.7	7.1	–0.9
Jammu and Kashmir					
	Chickpea	30.3	28.2	36.7	6.4
	Green gram	16.4	13.2	15.6	–0.8
	Black matpe	18.9	18.6	12.3	–6.6
	Lentil	3.9	4.1	5.1	1.2
Jharkhand	Chickpea	27.1	19.0	33.6	6.5
	Pigeon pea	28.6	28.9	23.2	–5.4
	Lentil	17.9	32.5	22.2	4.3
Karnataka	Pigeon pea	38.6	44.1	41.7	3.1
	Chickpea	14.9	15.7	19.6	4.7
	Green gram	12.3	11.1	10.8	–1.5
	Black matpe	7.4	9.9	11.0	3.6
Kerala	Chickpea	16.4	18.7	26.1	9.7
	Pigeon pea	21.3	19.9	18.6	–2.7
	Green gram	20.3	21.8	18.2	–2.1
Madhya Pradesh	Pigeon pea	34.9	42.0	39.8	4.9
	Chickpea	27.0	23.1	26.9	–0.1
	Green gram	12.7	11.7	13.9	1.2

(continued)

TABLE 2A.3 *(continued)*

State	Major pulses	1988 (%)	1999 (%)	2009 (%)	Change (%) (1988–2009)
Madhya Pradesh	Lentil	9.2	10.4	7.6	–1.6
	Black matpe	8.8	6.9	7.4	–1.4
Maharashtra	Pigeon pea	38.1	44.0	38.7	0.6
	Chickpea	25.6	22.4	26.8	1.2
	Green gram	14.3	15.2	15.5	1.2
	Lentil	5.3	6.6	6.0	0.7
	Black matpe	5.3	4.7	6.5	1.2
Manipur	Peas	23.3	26.9	24.9	1.6
	Lentil	15.9	20.4	17.0	1.1
	Chickpea	6.3	8.0	28.0	21.7
	Pigeon pea	12.7	3.0	4.4	–8.3
	Green gram	14.7	3.8	3.1	–11.6
	Black matpe	6.5	5.6	5.7	–0.8
Meghalaya	Lentil	75.3	76.8	58.4	–16.9
	Green gram	13.1	7.8	16.5	3.4
	Chickpea	3.9	4.3	14.6	10.7
Mizoram	Lentil	56.0	70.0	75.3	19.3
	Peas	12.6	8.2	2.9	–9.7
Nagaland	Lentil	77.0	54.4	42.1	–34.9
	Green gram	10.1	2.8	2.7	–7.4
	Chickpea	0.3	5.9	21.1	20.8
Odisha	Green gram	41.0	30.6	25.9	–15.1
	Pigeon pea	24.6	30.6	29.3	4.7
	Chickpea	4.4	8.6	16.2	11.8
	Lentil	9.1	13.4	5.2	–3.9
Punjab	Chickpea	34.4	38.3	48.4	14.0
	Green gram	22.9	22.1	19.0	–3.9
	Lentil	17.9	18.2	12.6	–5.3
	Black matpe	13.8	10.3	12.0	–1.8
Rajasthan	Chickpea	46.3	43.1	48.0	1.7
	Green gram	28.1	29.2	28.6	0.5
	Lentil	7.1	7.6	7.4	0.3
	Black matpe	9.0	11.4	7.4	–1.6
	Pigeon pea	3.9	5.1	5.4	1.5

State	Major pulses	1988 (%)	1999 (%)	2009 (%)	Change (%) (1988–2009)
Sikkim	Lentil	51.4	54.4	46.1	–5.3
	Green gram	11.6	5.6	3.2	–8.4
	Chickpea	3.5	10.7	15.3	11.8
Tamil Nadu	Pigeon pea	42.9	40.1	40.4	–2.5
	Black matpe	26.5	28.6	29.5	3.0
	Chickpea	12.7	12.9	13.5	0.8
	Green gram	7.6	8.8	6.9	–0.7
Tripura	Lentil	71.2	79.6	82.4	11.2
	Green gram	13.9	9.6	7.3	–6.6
Uttarakhand	Chickpea	16.8	22.9	33.0	16.2
	Pigeon pea	21.3	20.7	17.8	–3.5
	Lentil	22.6	22.6	14.0	–8.6
	Black matpe	17.3	12.1	10.7	–6.6
Uttar Pradesh	Pigeon pea	41.4	36.3	29.2	–12.2
	Chickpea	21.9	15.3	21.5	–0.4
	Black matpe	14.0	14.3	10.6	–3.4
	Lentil	11.0	13.1	8.3	–2.7
	Peas	5.3	15.1	24.6	19.3
West Bengal	Lentil	49.8	59.6	51.6	1.8
	Green gram	19.0	18.7	15.1	–3.9
All India					
	Pigeon pea	26.8	27.8	25.4	–1.4
	Chickpea	20.8	19.5	24.8	4.0
	Lentil	14.6	16.5	13.7	–0.9
	Green gram	13.6	13.4	12.1	–1.5
	Black matpe	10.7	10.8	11.0	0.3
	Peas	2.2	3.7	5.3	3.1

Source: Authors' calculations based on different NSS rounds.

Note: The shares of different pulses in states remains nearly same in the 68th round as in the 66th round.

TABLE 2A.4 Trends in protein intake from pulses across rural and urban households in Indian states, 1988–2009 (in grams per capita per day)

	Rural households			Urban households		
State	1988	2009	Change (%)	1988	2009	Change (%)
Arunachal Pradesh	4.0	3.9	−2.4	12.8	4.5	−64.4
	(4.7)	(6.0)		(15.6)	(5.9)	
Andhra Pradesh	6.3	5.2	−17.3	7.0	6.2	−10.9
	(11.2)	(7.7)		(13.6)	(8.0)	
Assam	5.9	4.2	−28.9	8.1	4.9	−39.6
	(10.9)	(7.0)		(14.5)	(7.2)	
Bihar	9.3	4.5	−52.0	8.5	4.9	−42.4
	(13.5)	(6.9)		(13.3)	(6.6)	
Delhi	8.7	4.6	−46.9	10.3	4.9	−51.8
	(9.9)	(7.7)		(14.4)	(5.4)	
Jharkhand	7.2	4.2	−41.8	8.5	5.3	−37.8
	(12.7)	(7.1)		(14.2)	(7.7)	
Gujarat	7.2	5.3	−25.9	8.9	6.1	−31.1
	(11.2)	(8.3)		(15.2)	(9.7)	
Haryana	6.8	4.6	−31.7	7.1	5.0	−29.4
	(7.6)	(5.7)		(10.5)	(5.6)	
Himachal Pradesh	10.4	9.1	−13.0	12.5	9.2	−26.7
	(12.7)	(10.3)		(15.9)	(10.5)	
Jammu and Kashmir	5.2	4.1	−20.8	4.7	4.1	−12.9
	(7.2)	(5.4)		(7.1)	(5.3)	
Karnataka	7.8	5.3	−32.6	8.0	5.9	−26.0
	(12.9)	(8.3)		(14.6)	(7.6)	
Kerala	4.0	4.6	13.5	4.7	4.8	2.3
	(7.3)	(6.3)		(8.8)	(6.4)	
Madhya Pradesh	10.2	5.5	−46.2	9.3	5.8	−37.0
	(13.7)	(8.3)		(14.2)	(8.4)	
Chhattisgarh	8.2	5.5	−33.7	10.3	6.9	−33.3
	(14.3)	(9.4)		(17.0)	(10.1)	
Maharashtra	8.6	6.6	−22.8	9.4	6.6	−30.1
	(13.4)	(9.8)		(15.5)	(8.4)	
Manipur	6.2	2.8	−54.2	6.7	2.5	−62.7
	(10.7)	(5.1)		(11.8)	(4.5)	
Mizoram	7.2	4.0	−44.9	10.3	4.7	−54.2
	(12.2)	(6.5)		(15.0)	(7.6)	

	Rural households			Urban households		
State	1988	2009	Change (%)	1988	2009	Change (%)
Nagaland	6.5	3.0	−54.1	6.8	3.3	−51.8
	(9.8)	(4.3)		(10.7)	(5.0)	
Odisha	4.5	4.4	−3.0	7.0	5.1	−26.3
	(8.2)	(7.4)		(11.8)	(7.3)	
Punjab	8.5	5.9	−29.9	8.2	6.3	−23.0
	(9.8)	(8.2)		(12.6)	(9.3)	
Rajasthan	5.2	3.1	−40.1	6.2	3.4	−43.9
	(5.9)	(4.2)		(8.6)	(4.9)	
Sikkim	4.5	3.6	−21.3	6.8	4.4	−34.9
	(8.8)	(5.5)		(12.2)	(7.4)	
Tamil Nadu	6.0	6.6	−10.2	7.4	7.0	−4.6
	(11.9)	(9.8)		(15.5)	(8.4)	
Tripura	4.8	3.5	−27.9	5.8	4.4	−22.8
	(8.3)	(4.9)		(10.0)	(5.7)	
Uttar Pradesh	9.7	5.7	−41.4	8.1	5.4	−32.9
	(12.5)	(8.2)		(12.2)	(8.0)	
Uttrakhand	8.8	6.1	−31.4	10.7	6.2	−41.5
	(11.4)	(8.6)		(16.0)	(8.4)	
West Bengal	4.3	3.4	−21.5	5.6	3.7	−33.7
	(7.5)	(5.3)		(9.9)	(5.1)	

Source: Authors' calculations based on different NSS rounds.

Note: Figures within the parentheses indicate share of pulses in total protein intake. Please see Table 2.1 for the 2011–2012 figures for protein intake from pulses.

TABLE 2A.5 Trends in protein intake from pulses by poor and rich households in Indian states, 1988–2009 (in grams per capita per day)

	Poor households			Rich households		
State	1988	2009	Change (%)	1988	2009	Change (%)
Arunachal Pradesh	2.4	2.3	−4.4	6.8	5.1	−25.1
	(5.4)	(5.7)		(6.6)	(6.0)	
Andhra Pradesh	3.6	3.2	−10.9	9.2	6.9	−24.4
	(9.0)	(7.5)		(13.7)	(8.0)	
Assam	4.1	2.6	−36.5	9.4	5.9	−36.9
	(9.5)	(6.0)		(13.8)	(7.7)	
Bihar	6.8	3.4	−50.8	14.5	6.2	−57.1
	(12.2)	(6.4)		(15.6)	(7.2)	
Chhattisgarh	5.5	3.6	−33.9	14.5	7.6	−47.4
	(12.0)	(8.2)		(18.5)	(10.4)	
Delhi	5.7	3.8	−32.7	11.6	5.2	−55.3
	(11.7)	(9.4)		(14.2)	(5.0)	
Jharkhand	4.5	3.0	−33.5	13.0	6.3	−51.5
	(9.8)	(6.7)		(16.8)	(7.8)	
Gujarat	5.9	3.9	−34.1	10.4	7.6	−27.4
	(11.8)	(8.1)		(13.6)	(9.6)	
Haryana	4.0	3.1	−22.3	8.5	5.5	−35.4
	(7.4)	(5.6)		(8.8)	(5.6)	
Himachal Pradesh	6.7	5.8	−13.9	13.2	10.0	−24.3
	(11.1)	(9.9)		(14.2)	(10.2)	
Jammu and Kashmir	3.9	2.9	−24.0	6.3	4.7	−24.7
	(6.9)	(5.7)		(7.6)	(5.4)	
Karnataka	5.2	4.0	−23.3	11.2	6.5	−42.1
	(11.8)	(9.1)		(14.9)	(7.3)	
Kerala	1.6	2.0	22.5	7.0	5.6	−19.9
	(4.5)	(5.6)		(9.5)	(6.4)	
Madhya Pradesh	6.6	4.2	−37.5	14.4	6.9	−51.9
	(11.8)	(8.5)		(15.7)	(8.5)	
Maharashtra	6.6	4.5	−31.6	11.8	7.6	−35.1
	(12.8)	(9.2)		(15.5)	(8.9)	
Manipur	3.6	2.1	−42.8	8.1	3.3	−58.5
	(8.2)	(4.7)		(12.5)	(4.8)	
Meghalaya	2.0	1.5	−23.7	5.3	2.4	−55.2
	(4.3)	(3.4)		(9.0)	(3.5)	

	Poor households			Rich households		
State	1988	2009	Change (%)	1988	2009	Change (%)
Mizoram	3.7	2.1	−42.1	9.0	5.2	−42.4
	(8.4)	(5.1)		(13.7)	(7.6)	
Nagaland	1.7	2.0	22.1	6.9	3.6	−48.2
	(4.4)	(4.0)		(10.8)	(4.5)	
Odisha	2.8	2.8	0.2	9.9	6.0	−39.1
	(6.2)	(6.5)		(12.7)	(7.8)	
Punjab	5.3	4.3	20.1	9.8	6.9	−29.7
	(10.4)	(9.3)		(10.8)	(8.5)	
Rajasthan	3.0	2.0	−31.0	8.3	3.9	−52.9
	(4.7)	(3.8)		(8.2)	(4.7)	
Sikkim	3.9	2.3	−41.5	7.1	4.2	−40.9
	(9.3)	(4.7)		(10.9)	(5.8)	
Tamil Nadu	3.5	4.4	24.1	11.2	7.8	−30.6
	(9.7)	(10.9)		(16.6)	(8.5)	
Tripura	3.0	2.3	−23.4	6.3	4.3	−31.9
	(7.1)	(4.8)		(9.3)	5.3	
Uttrakhand	5.9	4.3	−26.9	11.4	7.1	−37.2
	(10.0)	(9.0)		(13.1)	(8.3)	
Uttar Pradesh	6.5	4.2	−35.8	13.2	7.1	−46.6
	(10.8)	(7.7)		(13.9)	(8.5)	
West Bengal	3.3	2.2	−33.0	6.9	4.3	−37.0
	(7.3)	(5.2)		(9.2)	(5.1)	

Source: Authors' calculations based on different NSS rounds.

Note: Figures within the parentheses indicate share of pulses in total protein intake.

Chapter 3

TEMPORAL AND SPATIAL DYNAMICS OF PULSE PRODUCTION IN INDIA

Inba Sekar, Devesh Roy, and P. K. Joshi

Starting in the 1960s, the innovations of the "Green Revolution" brought about major benefits to the farming of rice and wheat in India, but the benefits were unevenly distributed across regions, and they had an unintended but generally negative impact on the pulses sector. The details of this trend are among the production dynamics we examine in this chapter for the entire pulses sector across regions and over time. The Green Revolution benefited rice and wheat farmers in the irrigated zones in India's northwest region, but failed to help farmers in other regions where farmers must depend on rainfall. Moreover, the expansion of rice and wheat farming had a deleterious impact on several other crops, including pulses, largely replacing them in the planting of acreage due to the new, easy availability of high-yielding varieties and access to irrigation. As they were displaced from their former traditional regions, pulses also experienced several changes both in regional specialization and in the adoption of technology. Altogether, during the height of the Green Revolution (between 1960–1961 and 1980–1981) pulse production declined from 12.6 million metric tons to 10.5 million metric tons. Looking at recent statistics, out of the total irrigated area in India in 2010, only 12 percent was planted in pulses while more than 60 percent was planted in wheat and rice paddy.[1]

Background

Over time, the growth in production in irrigated areas has gone up for all crops but for pulses at a rate lower than other crops. Based on Ministry of Agriculture data, in 1950–1951, 9.4 percent of the area of pulses was under irrigation. It came down to 8 percent by 1960–1961, with an increase to 8.8 percent share in 1970–1971, rising to 12.5 percent by 2000–2001. It peaked to 16.2 percent in 2007–2008, only to fall to 14.8 percent by

1 Hereafter, "tons" refers to "metric tons."

2010–2011. In this context, the case of pigeon pea is particularly stark. While in 1990–1991, 5.3 percent of the area was under irrigation, it came down heavily to 3.9 percent by 2011–2012. In contrast, based on the same statistics, while only 18 percent of the area under food grains had irrigation in 1950–1951, nearly half of the area under food grains had irrigation by 2011–2012 (India, Ministry of Agriculture 2014). Today, the productivity of pulse farming in India, which averages 694 kilograms per hectare, continues to lag behind that recorded in most of the other major pulse-producing countries. It also lags behind the yield attained at Indian research stations and on-farm demonstrations (Reddy, Bantilan, and Mohan 2013). Studies show that since pulses were pushed by high-yielding cereal crops to marginal environments, the area put into pulse planting has been largely determined by rainfall. Conversely, the availability of irrigation coverage generally reduces the acreage planted under pulses (Sadavatti 2007).

The returns to pulse farmers are also a challenge. Although the harvest prices of pulses are usually much higher than those received for competing crops, especially rice and wheat, farmers still do not realize reasonable returns because pulse yields are lower and more unstable. Moreover, the government's pricing policy and procurement support for rice, wheat, cotton, and sugarcane have likely adversely affected the farmers' decision to grow pulses. This is because the government carries only negligible procurement for pulses, unlike its procurement for the other crops, so support prices in pulses are only notional and fall well below the market prices (at the retail level). However, note that the implementability and likely effectiveness of a large procurement program for pulses cannot be taken for granted and warrant further study.

Also, farmgate prices are too low to cover the risk premium associated with pulse production, thereby limiting the supply response of farmers. Accessing credit to support their pulse farming is a challenge: in 2001, disbursements to farmers were 85 rupees per hectare for pulses as compared with 458 rupees per hectare for rice and 90 rupees per hectare for wheat (Materne and Reddy 2007; Reddy 2009b). Other factors besides price play important roles in determining the acreage and production of different pulses (Tuteja 2006). In allocating land to different pulses, lagged acreage and magnitude of presowing rainfall (and residual moisture particularly for *rabi* pulses) are important considerations. Pulse crops are susceptible to many biotic and abiotic stresses due to indeterminate plant type, which makes the development of biotic and abiotic stress-resistant varieties difficult to develop (Reddy 2009b). Srivastava, Sivaramane, and Mathur (2010) argue that the demand-driven and location-specific improved varieties and technological options are more limited for pulses than they are for rice and wheat. Overall, there have been few significant technological breakthroughs for pulses due to peculiar problems like indeterminate plant type, low response to fertilizers, and management practices (Reddy, Bantilan, and Mohan 2013). Furthermore, research and development (R&D) for pulses has received less attention and funding than that for other crops by both international organizations and multinational corporations. These issues are discussed further in Chapter 4.

The spatial movement of pulse cultivation across years is one of the most important determinants of the crop's supply-side dynamics. Currently, just six states account for 80 percent of India's total pulse production: Andhra Pradesh, Karnataka, Madhya Pradesh, Maharashtra, Rajasthan, and Uttar Pradesh (India, Ministry of Agriculture and Cooperation 2010). Several studies show that pulses production in India is finding new niche areas where it is reportedly performing better than in the traditional ones (Joshi, Asokan, and Bantilan 1999; Shiyani et al. 2000; Joshi and Saxena 2002). Finally, notwithstanding the marginalization of pulses over the years, the government has recently been proactive in promoting pulses through various schemes. The external market environment has changed significantly over the course of time, with increased import penetration in different pulses and the emergence of some pulse exports from India. Indeed, India achieved a record output in pulse production at 18.1 million tons in 2010–2011, which included an all-time high production in chickpea (8.25 million tons), green gram (1.82 million tons), and black matpe (1.74 million tons). Despite the significant and dynamic changes occurring across space and over time, research on the evolution and performance of the pulse sector's supply side in India remains scant. This chapter attempts to fill the research gap.

Study Objectives and Methods

The study presented here aims (1) to identify the spatial and temporal variations in pulse production in India, with special emphasis on chickpea and pigeon pea, and (2) to identify the factors affecting pulses' relative area allocation across the country.

Definition of Periods

To study temporal variations in pulse production, we divided the study period into the following four periods:

1. **Pre- and initial phase of Green Revolution (1960–1970).** This decade was a period of stagnation in both the area devoted to pulses and their total production. There was no major breakthrough in pulse research and no government scheme was launched to support pulses production in the country.

2. **Mature phase of Green Revolution (1971–1990).** During this period, both area cultivated and production of pulses decreased in the country's northern and eastern zones and increased in the southern and western zones. Several pulse development schemes were launched during this period, including:

    - Pulses Development Scheme (Fourth Five-Year Plan) (1969–1970 to 1973–1974)
    - National Pulses Development Project (Seventh Five-Year Plan) (1985–1986 to 1989–1990)
    - Special Food Grain Production Program (1988–1989)

3. **Post-liberalization period (1991–2000).** This period witnessed a further increase in pulse area and production in the southern and western states. Favorable terms of trade were reached for agriculture relative to industry, which may have influenced cropping patterns. Some short-duration and wilt-resistant varieties of pulses were developed during this period.

4. **Post-trade spike period (2001 and beyond).** This period witnessed a third spike in pulse area and production in the southern and western states. The government launched schemes for pulse development including:

    - Integrated Scheme of Oilseeds, Pulses, Oil Palm and Maize (ISOPOM) (2004)
    - National Food Security Mission (NFSM) (2007–2008)
    - Special plan to achieve more than 19 million tons of pulse production by *kharif* (the rainy season) (2012–2013)

These efforts could raise pulses production to an extent culminating in production of 18 million tons in 2011–2012, but pulses imports also increased post-2001, implying that the increase in production of pulses was still not sufficient to meet their rising domestic demand.

Definition of Zones

To study the spatial movement (the geographic dispersion) of pulse cultivation over time, we grouped the states into five zones based on their geographical location. The *northern zone* comprises Haryana, Himachal Pradesh, Punjab, Jammu and Kashmir, and Uttar Pradesh. The *southern zone* includes the states of Andhra Pradesh, Karnataka, and Tamil Nadu. The *eastern zone* consists of the states of Assam, Bihar, Odisha, and West Bengal, while the *western zone* includes Rajasthan, Gujarat, and Maharashtra. The *central zone* has only two states, Madhya Pradesh and Chhattisgarh.

An extensive mapping of the movement of pulse area and production across space and over time allowed us to cluster different states into specific groups. These clusters depict the status of pulse cultivation and production. The states that showed secular upward trends in production were labeled "gaining-ground states"; those showing secular downward trends were labeled "losing-ground states"; and those showing in-between patterns were labeled as "status quo states." The analysis was disaggregated by crops, with a focus on two major pulses: chickpea and pigeon pea. These two pulses currently constitute 60 percent of all pulses in India both by area cultivated and by production. Through a state-level analysis, we studied the spatial dynamics of these pulses in response to factors such as the Green Revolution, the economic reforms of 1991, and pulse trade (import) spikes since 2000.

Comparative Performance: Pulses versus Cereals

Table 3.1 summarizes the relative performance of pulses as compared with cereals between 1960 and 2010 in terms of area, production, and yields.

First Period: 1960–1970

During this period the total pulse crop area in India declined by around 8 percent and, concomitantly, the area and production of both wheat and rice gained momentum. The area and production of chickpea declined by 21.64 percent and 9.73 percent, respectively, while the area and production of pigeon pea increased by 8.26 percent and 10.68 percent, respectively. The yield of chickpea increased by about 15 percent, but the yield of pigeon pea increased only marginally. The yield of all pulses, in total, increased by only 8 percent—clearly a subpar performance relative to cereals, such as wheat, which rose in yield by 55 percent.

TABLE 3.1 Comparative performance of cereals and pulses, 1960–2010 (% change)

Crop	Particulars	1. Pre– and initial phase of Green Revolution (1960–1970)	2. Matured phase of Green Revolution (1971–1990)	3. Postliberalization period (1991–2000)	4. Post–trade spike period (2001–2010)[a]
Pulses	Area	−8.02	3.31	−0.93	2.25
	Production	−0.87	10.34	2.15	14.08
	Yield	7.97	6.72	3.67	11.23
Wheat	Area	26.11	48.50	15.62	3.18
	Production	95.54	171.50	42.55	12.13
	Yield	55.13	83.00	23.33	8.73
Paddy/rice	Area	11.88	10.50	8.72	−1.53
	Production	33.77	70.60	28.55	10.34
	Yield	19.72	54.00	18.54	12.09

Source: Authors' calculations based on data from India, Ministry of Agriculture, Directorate of Economics and Statistics, 1950–2010.

Note: a. Although in recent years pulses have fared better, in a comparative sense they still lag behind cereals starting from a low base.

Second Period: 1971–1990

Despite implementation of the government's Pulses Development Scheme during this period (the Fourth Five-Year Plan), the total area, production, and yield of pulses increased by only 3.3 percent, 10.3 percent, and 6.7 percent, respectively. In comparison, for wheat the increase in area, production, and yield was 48 percent, 171 percent, and 83 percent, respectively; for rice, it was 10 percent, 70 percent, and 54 percent, respectively. The chickpea yield increment was merely 0.5 percent. Consequently, the 18 percent decrease in chickpea area was accompanied by an 18 percent decrease in its production. In contrast, pigeon pea increased in both area and production, and its yield showed an increase of more than 7 percent during this period.

Third Period: 1991–2000

During the third period, the total pulse area declined, whereas the area under wheat and rice increased by 15 percent and 9 percent, respectively. The yields of both wheat and rice also increased. While pigeon pea showed a marginal decline, chickpea enjoyed a remarkable increase in area, production, and yield of nearly 18 percent, 39 percent, and 19 percent, respectively. The introduction of short-duration chickpea varieties in Andhra Pradesh and Karnataka during the early 1990s might explain these significant increases.

Fourth Period: 2001 and Beyond

After 2001, there was a spike in the trade of pulses, with a 36 percent increase in imports, while at the same time chickpea, pigeon pea, wheat, and rice all showed a rise in production and yield. Consequently, there seems to be little evidence to support the idea of pulses imports crowding out domestic production. Importantly, during this period, for the first time, pulses' yield increased by more than 11 percent. The success of the Technology Mission on Oilseeds (TMO) induced the government to restructure the program in 2004 by adding pulses and maize to create the Integrated Scheme of Oilseeds, Pulses, Oil Palm, and Maize.[2] Historically, India had been a net importer of edible oils. After a period of stagnation in oilseed production and large imports, the TMO was introduced in 1986, which increased oilseed production significantly. This is now widely known as the "yellow revolution" (Reddy 2009a).

Under ISOPOM, financial assistance was provided for the production and distribution of certified seeds, seed mini-kits, sprinkler sets, *Rhizobium* culture, and phosphate solubilizing bacteria (PSB), gypsum/pyrite, plant protection chemicals, and biofertilizers. Demonstrations of integrated pest management (IPM) were organized on a large scale through the State Department of Agriculture. To address the shortage of quality seeds, a provision of credit to produce pulse seeds was also included in the scheme. In addition, in 2008 the Accelerated Pulses Production program was launched under the National Food Security Mission and implemented in 468 districts in 16 states.

Changing Composition of Pulse Production

Since 1991, the relative share of total pulse production made up by chickpea has risen, that made up by pigeon pea has declined, and the contribution of minor pulses like lentil and peas has remained nearly the same. Table 3.2 depicts this varying composition of the pulse sector in national production. During the triennium ending (TE) 1991, chickpea contributed 36 percent to total pulses production, and by TE 2012 its contribution had risen to 47 percent.

2 The Technology Mission on Oilseeds was launched by the central government in 1986 to increase India's production of oilseeds to reduce import and achieve self-sufficiency in edible oils.

TABLE 3.2 Composition of pulses production (share of total production, %)

Pulse	Triennium ending 1991	Triennium ending 2000	Triennium ending 2012
Chickpea	36	42	47
Pigeon pea	19	17	16
Lentil	5	7	6
Black matpe	12	9	10
Green gram	10	7	8
Peas	4	5	4
Others	14	13	9

Source: Authors' calculations based on data from India, Ministry of Agriculture and Cooperation,1950–2010.

Case Studies of Chickpea and Pigeon Pea

The changing behavior of two key pulse crops—chickpea and pigeon pea—over the four periods and across different zones (and states within zones) is presented next.

Chickpea

Area. The temporal changes depicted in Table 3.3 reveal that at the country level, chickpea area decreased during the first two periods (pre– and mature Green Revolution) and then increased during the latter two periods (postliberalization and post–trade spike). Underlying these changes are significant regional differences in production and yield. In the northern zone, which is the principal area for chickpea consumption, the chickpea cultivation area declined continuously from the first period on. This represents a salient feature of the transition in pulses, during which the location of production and that of consumption became separated over time.

The introduction of the chickpea crop into nontraditional areas, including the southern Indian states, is an example of a technological and institutional breakthrough (Reddy, Bantilan, and Mohan 2013). During 1991–1993 to 2006–2008, the highest increase in chickpea productivity was recorded in Andhra Pradesh (124 percent), followed by Karnataka (63 percent), Maharashtra (52 percent), and Gujarat (40 percent). The factors usually considered most responsible for this expansion of chickpea into southern India are the introduction of chickpea into black cotton soils, the availability of plenty of *rabi* fallow land, and the adoption of short-duration and high-yielding varieties.

TABLE 3.3 Decadal change in chickpea area, 1960–2010 (%)

State	First period: Pre– and initial phase of Green Revolution (1960–1970)	Second period: Mature phase of Green Revolution (1971–1990)	Third period: Postliberalization (1991–2000)	Fourth period: Post–trade spike (2001–2010)
Northern Zone				
Haryana	−34.1	−54.5	−40.7	−61.3
Himachal Pradesh	−47.7	−59.3	−64.9	−56.7
Punjab	−53.1	−85.4	−83.0	−73.4
Uttar Pradesh	−15.2	−40.6	−35.5	−35.2
Southern Zone				
Andhra Pradesh	−24.2	−33.5	189.7	313.1
Karnataka	34.2	4.4	54.5	127.5
Tamil Nadu	37.5	64.5	33.1	−12.4
Eastern Zone				
Assam	—	73.3	−23.0	−31.2
Bihar	−50.0	−38.0	−34.3	−42.9
Odisha	19.3	98.0	−26.8	23.3
West Bengal	−14.5	−74.4	−41.0	−9.0
Western Zone				
Gujarat	−65.7	95.6	42.7	54.5
Maharashtra	−13.1	60.6	37.5	48.4
Rajasthan	−20.7	−13.6	93.1	−43.7
Central Zone				
Madhya Pradesh	3.1	34.4	21.6	6.9
All India	−21.6	−18.4	17.6	6.4

Source: Authors' calculations based on India, Ministry of Agriculture 2015.

Note: — = data not available. Data on Madhya Pradesh includes Chhattisgarh. We also presented the area allocation and production of pulses across different states for chickpea and pigeon pea in maps (Figure 3A.3 and 3A.4) in the chapter appendix.

The chickpea variety known as KAK-2, a Kabuli type with higher market demand, was one example of the newly introduced varieties. Short-duration and wilt-resistant varieties like JG-11 were also influential. Stable yield and prices and a well-developed land-lease market, which facilitated large-scale mechanization, also contributed to the expansion of chickpea in southern India. The wider availability of highly subsidized cold-storage warehouses helped farmers to store chickpea during the peak harvest season to overcome lower market prices and to reap profits from higher prices during later periods (Reddy, Bantilan, and Mohan 2013).

During the fourth period studied—the post–trade spike decade after 2001—chickpea area increased quite significantly in the southern zone (Table 3.3). It increased by 313 percent in Andhra Pradesh and 128 percent in Karnataka (although another southern state, Tamil Nadu, experienced a decrease in chickpea area during the same period). Gujarat and Maharashtra are the two states in western India that have shown continuous increase in chickpea area from the mature phase of the Green Revolution period on. In the eastern region, Odisha showed a roughly 23 percent expansion in chickpea area, resuming a trend from a decade earlier during the mature phase of the Green Revolution, when both Odisha and Assam had increases.

The causes of these expansions in chickpea are known. In addition to the introduction of new technology, chickpea, which is less labor-intensive, was substituted for more labor-intensive preceding crops, including cotton, chilies, and other cash crops. Risk consideration also factored in. Crops such as cotton are prone to pests and diseases and their prices are subject to high fluctuations. Given the varied dry-land agroclimatic conditions of Andhra Pradesh, chickpea, a comparatively low-risk crop, became farmers' preferred alternative. It has lower propensity to damage by pests and disease than many other crops, it has better storability, and it suffers less from price fluctuations (Suhasini et al. 2009). In contrast, in most of the northern and eastern states, the trend in chickpea area was a declining one. This might be attributable to the region's extensive irrigation facilities, the nonavailability of high-yielding varieties of pulses, and an environment that has generally been favorable for rice and wheat cultivation through minimum support prices and assured procurement.

Production. At the country level, total chickpea production declined during both the initial and the matured periods of the Green Revolution (1960–1990), but then increased in the succeeding two decades—by 39 percent during the postliberalization period and by 12 percent during the post–trade spike (see Table 3A.2 in the chapter appendix for detailed production figures). In the northern zone, all states showed a decreasing trend in chickpea production, irrespective of period, which is mainly attributable to chickpea's substitution there by cereals.

In the southern zone, the growth in chickpea production in both Andhra Pradesh and Karnataka has been quite remarkable. During the postliberalization and post–trade spike decades it rose in Andhra Pradesh first by 288 percent and then by 730 percent, while in Karnataka it rose over the same two decades by 106 percent and then 172 percent. In the central zone, Madhya Pradesh had substantial increases during both the matured phase of the

Green Revolution (75 percent) and the postliberalization period (67 percent); although the increases slowed to a little more than 5 percent during the recent post–trade spike period, the earlier sustained growth resulted in the state being a major producing area (Table 3A.2).

In the western zone, both Maharashtra and Gujarat have shown continuous if somewhat erratic increases in chickpea production from the matured phase of the Green Revolution period on through the postliberalization and post–trade spike decades. During these three periods, production in Gujarat sequentially rose by 28 percent, 94 percent, and 92 percent. In Maharashtra it rose sequentially by 187 percent, 50 percent, and 107 percent. The variation in these increases was due to increased price variability, more erratic rainfall patterns, and fluctuations in the supply of modern inputs like pesticides (Reddy 2006). The important factors that influenced the adoption of improved chickpea varieties in Gujarat included the short duration of crop, suitable farm size, lowered yield risk, and considerable earlier experience of growing chickpea crop (Shiyani et al. 2000).

Yield. The yield patterns in chickpea across regions are presented in Figure 3.1. In both northern and southern zones, the yields increased significantly in some states during the postliberalization period. In Andhra Pradesh the spike in yields was particularly striking in the previous decade, the period before the trade spike. Neither the eastern nor the western zone fared well in yield, however. Madhya Pradesh experienced a significant yield increase of more than 37 percent during the postliberalization period, which, owing to high base effect, culminated in small growth during the post–trade spike period.

Pigeon Pea

Area. During the Green Revolution, the total area planted in pigeon pea in India increased significantly (32.5 percent). It decreased slightly during the postliberalization period (by 1.8 percent) and rose again slightly (by 3.3 percent) in the post–trade spike period (Table 3.4). This is illustrated in an extreme form by the changes in Haryana. During the matured phase of the Green Revolution, Haryana showed a remarkable increase of more than 400 percent in pigeon pea area, although that was followed by two decades of slight decline. The initial spurt in Haryana is considered to be a function of the low base and the advent of short-duration varieties that fit well in the pigeon pea wheat cropping mix (Singh et al. 1996). Overall, despite Haryana's initial spike, the northern zone has shown a decreasing trend in pigeon pea area since the postliberalization period.

FIGURE 3.1 Chickpea yield variations in various zones (% change)

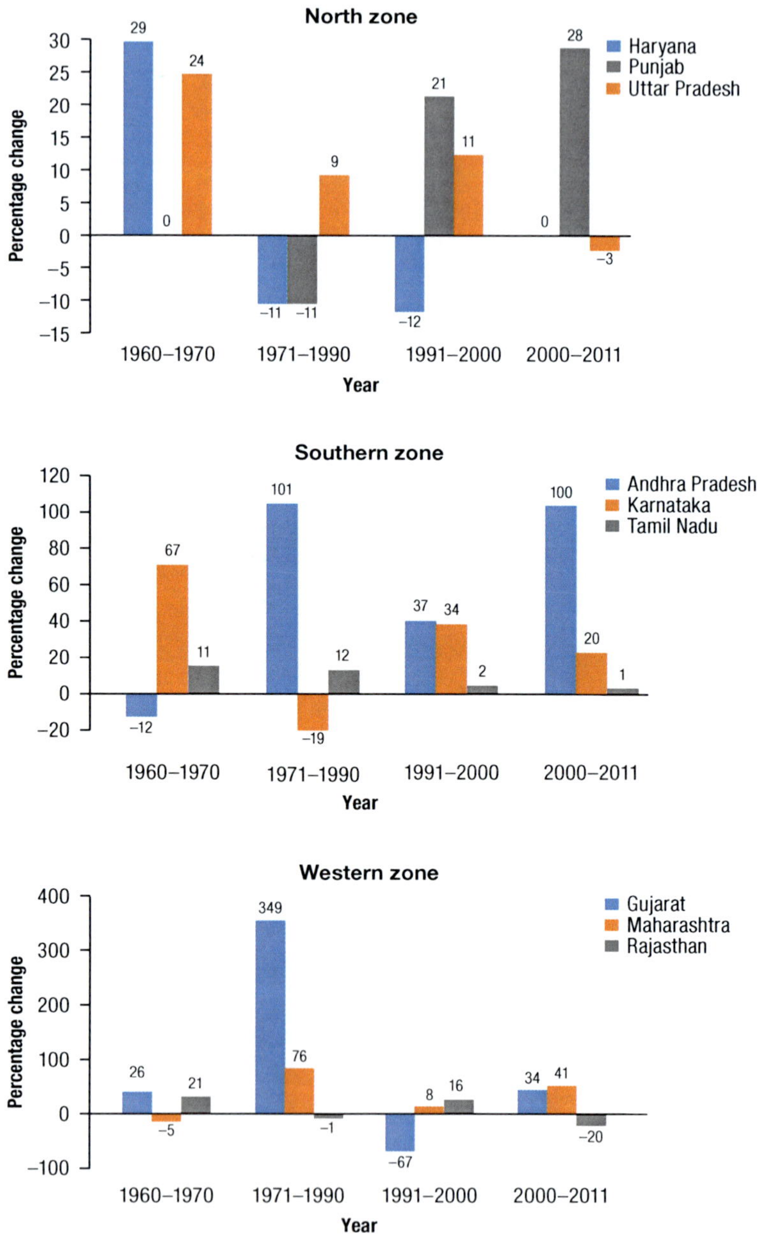

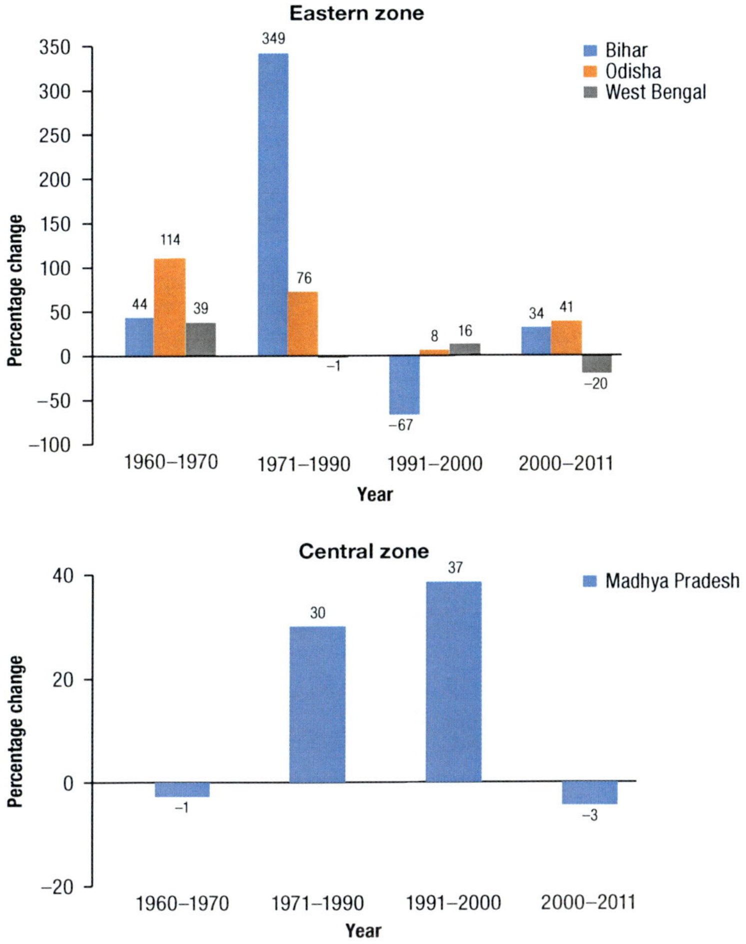

Source: India, Ministry of Agriculture, Directorate of Economics and Statistics, 1960–2010.

Two states that experienced continuous increases in pigeon pea area were Andhra Pradesh in the south and Maharashtra in the west. Gujarat, in the western zone, also witnessed a significant increase in both production and area during the matured phase of the Green Revolution, followed by a decline during the next two decades. In the central zone, Madhya Pradesh had a continuous decrease in pigeon pea area; the decline equaled 19.8 percent during the postliberalization period and 6.7 percent during the post–trade spike period.

Production. At the all-India level, pigeon pea production increased by 10.6 percent and 42.6 percent, respectively, during the initial and matured phases of the Green Revolution (mainly during the 1980s). It then declined by 5.8 percent during the postliberalization period and increased again

TABLE 3.4 Decadal change in pigeon pea area, 1960–2012 (%)

State	Pre– and initial phase of Green Revolution (1960–1970)	Matured phase of Green Revolution (1971–1990)	Postliberalization period (1991–2000)	Post–trade spike period (2001–2012)
Northern Zone				
Haryana	—	411.6	−30.1	−5.9
Punjab	—	—	−60.5	−41.7
Uttar Pradesh	−5.3	−17.6	−13.0	−24.4
Southern Zone				
Andhra Pradesh	12.1	93.4	12.1	19.5
Karnataka	−1.9	70.2	−2.5	33.9
Tamil Nadu	−6.5	187.7	−48.8	−63.0
Eastern Zone				
Assam	83.3	119.0	−11.6	−17.3
Bihar	−7.8	−59.9	−7.4	−54.2
Odisha	163.4	307.3	−4.0	−3.1
West Bengal	26.8	−83.5	−45.6	−73.1
Western Zone				
Gujarat	15.9	276.9	6.6	−28.4
Maharashtra	11.9	41.9	16.1	7.3
Rajasthan	11.2	4.4	33.7	−42.6
Central Zone				
Madhya Pradesh	30.6	−9.6	−19.8	−6.7
All India	8.2	32.5	−1.8	3.3

Source: India, Ministry of Agriculture, 1950–2010.
Note: — = data not available.

by 7.6 percent during the post–trade spike period. In the northern zone, Haryana, after achieving a remarkable increase of 853 percent in production during the matured phase of the Green Revolution, showed a decline subsequently. In the southern zone, Karnataka has shown a continuous rise in pigeon pea production and has, in fact, become the leading state in pigeon pea production in recent years. In the past decade, pigeon pea production in Karnataka went up by as much as 77 percent.

Andhra Pradesh, which rose in the ranks of chickpea production, has also shown a significant rise in pigeon pea production during the postliberalization and post–trade spike periods, after experiencing a decline of 25 percent during the matured phase of the Green Revolution. In the eastern zone, Odisha and

West Bengal had production increases of 40 percent and 68 percent, respectively, during the post–trade spike period. In the western zone, Gujarat and Rajasthan experienced declines in the post–trade spike period, but Maharashtra experienced a continuous rise from the mature phase of the Green Revolution on.

Yield. In general, there has been little net change in pigeon pea yield in India over the course of the entire period 1970–2010 (Figure 3.2 plots the evolution of pigeon pea yields across states over time). At the all-India level, there was a 4 percent decrease in pigeon pea yield during the postliberalization period. In the southern zone, although Andhra Pradesh and Karnataka showed a fall of 61.2 percent and 10.3 percent, respectively, in pigeon pea yield during the matured phase of the Green Revolution, they reversed that trend during the postliberalization and post–trade spike periods. (Source: India, Ministry of Agriculture, Directorate of Economics and Statistics)

Dynamics of Pulses across States

Based on the above findings concerning the area allocation and production of pulses, we created a typology of states to identify their patterns, listing them under three clusters: states that gained ground, states that lost ground, and status quo states. This typology is presented in the chapter appendix in Figure 3A.5 and Figure 3A.6 for chickpea and pigeon pea and in Table 3A.4 for total pulses. What those figures show is that the number of gaining-ground states was far lower than the number of losing-ground states, resulting in an overall decline in the prominence of pulses. This picture emerges at the country level from Table 3A.4.

For both chickpea and pigeon pea, Andhra Pradesh, Karnataka, Madhya Pradesh, and Maharashtra are the only states that could be regarded as gaining-ground states. Importantly, no northern-zone states are included in the gaining-ground group. In fact, most of the losing-ground states are from the northern and eastern zones. The status quo states, which had either fluctuating (with no secular trend) or stagnant change in area and production for chickpea over time are Gujarat, Rajasthan, and Tamil Nadu. (No state could be classified as status quo state for pigeon pea.) Figure 3A.1 and Figure 3A.2 in the chapter appendix show the shifts in pulse-producing areas across states and districts for the pulse-producing states. The mapping of these shifts in the top pulse-producing states shows that pulse cultivation has mostly moved to the marginal districts. This is clearly true in the case of chickpea and pigeon pea.

FIGURE 3.2 Variation in pigeon pea yield across different zones and states of India (% change)

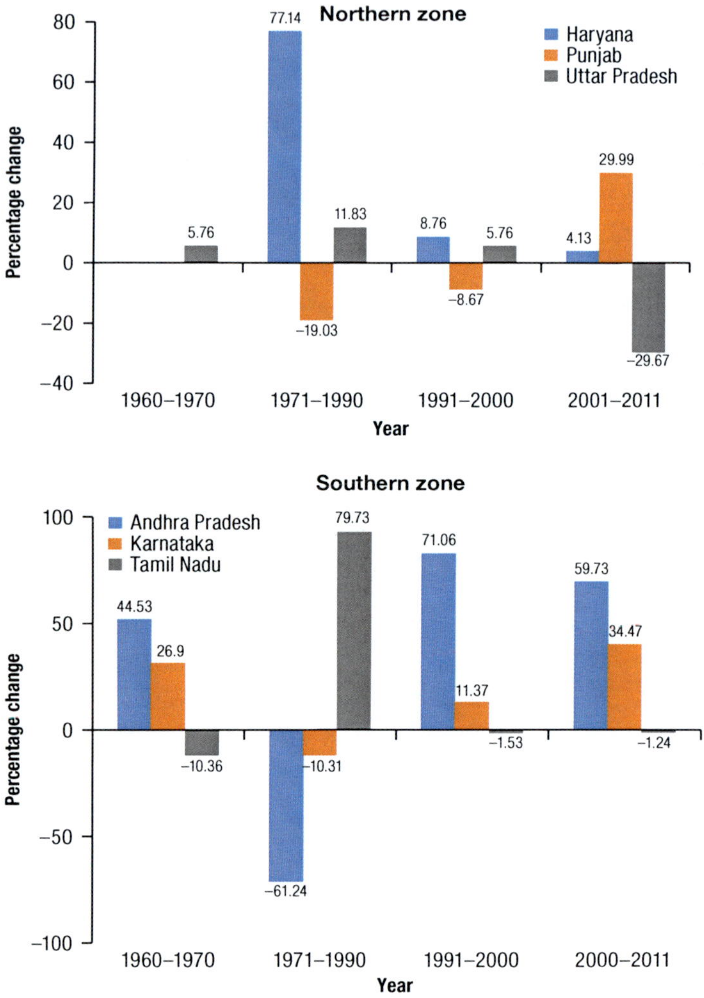

Determinants of Area Allocation and the Production of Pulses

Despite significant efforts by the government, as the discussion in this chapter shows, pulse production and productivity in India have been largely stagnant. Total production increased only by about 47 percent over the five decades reviewed. Specifically, it rose from about 12.5 million tons in the triennium ending 1960–1961 to about 18.5 million tons in the triennium ending

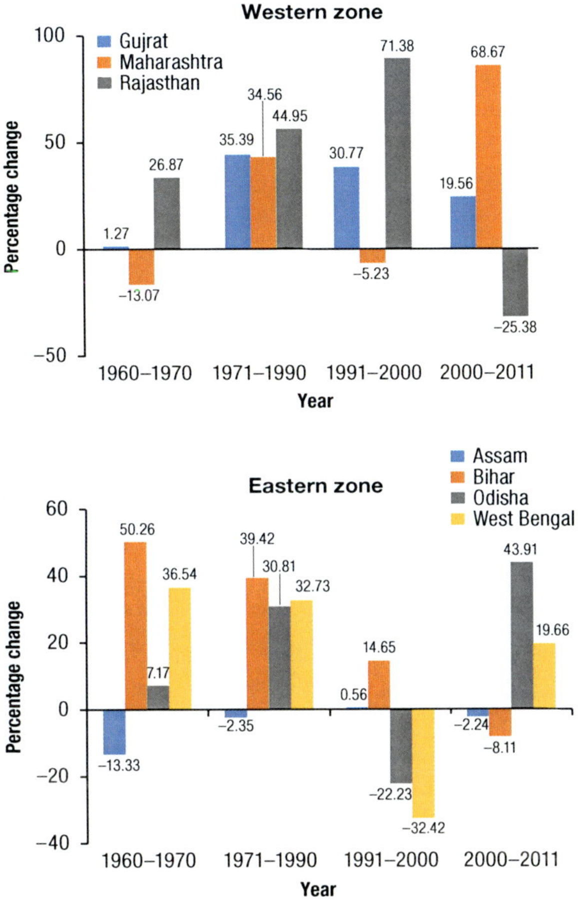

Source: India, Ministry of Agriculture, Directorate of Economics and Statistics.

2013–2014. By comparison, the production of rice and wheat have gone up over the same period—rising for rice by more than 225 percent to 106 million tons and for wheat by 808 percent to 95 million tons.

The persistent demand-supply gap in pulses is likely to widen if domestic production is not raised substantially. However, this is likely to be challenging since, according to Reddy (2009b), the low-productivity and low-input nature of pulse crops means they are generally grown as residual or alternate crops on marginal lands. This means most farmers plant pulses only after they have taken care of their food and income needs by planting high-productivity, high-input crops like rice and wheat. Pulses are overwhelmingly grown under rainfed conditions with little or no modern yield-enhancing inputs. On account of these factors, the supply of pulses has been constant for a long period (for more than three decades) and has increased to an average of 17 million tons only recently.

There are several reasons for the pattern of slow growth in pulse production. First, there has been continuous substitution of other crops for pulses. Competing crops (particularly rice and wheat) are more profitable under irrigated and adequate rainfall environments. Hence, incentives dictate that in these environments farmers opt for crops other than pulses. Not only have pulses been replaced by other crops in favorable conditions, a shift in pulse cultivation has been to less productive marginal drylands—precisely the areas where competing crops would not be worthwhile (Lingareddy 2015). The cropping in marginal environments without irrigation and the inherent susceptibility to pests end up making pulses both a low yield and a comparatively risky crop (Chand 2008). This, in turn, discourages farmers from making investments or growing pulses intensively with better inputs. Development of technology to increasingly cater to the marginal environments while improving yields and their stability has been difficult. This is one of the principal reasons for which aggregate supply of pulses has been comparatively inelastic. Further discussion of supply response is provided in the next section.

According to Reddy, Bantilan, and Mohan (2013), there are four additional reasons for the inelastic supply of pulses. First, the various types of pulse crops are scattered and thinly distributed, cultivated mostly in marginal and low-productivity lands, with each crop contributing a small share in total pulses area. These factors are a major hurdle for all stakeholders, including researchers and extension, development, and credit/market support agencies in offering both public- and private-sector input and output services and other institutional support. A second reason is the indeterminate plant type of many pulse crops combined with their low yield potential. Pulse crops are affected significantly by different pests and diseases during the crop-growing stage and also after harvest, with losses estimated by most studies to be in the range of 15 percent to 20 percent (IIPR 2011). Third, pulses show a low response to input management. Fourth, policy makers have accorded pulses a low priority (Materne and Reddy 2007).

The Conundrum of Rising Prices and Stagnant Production

Because the production of pulses has been subpar, the prices of major pulses such as pigeon pea and black matpe in their split forms (dals) have been increasing significantly, trading above 100 rupees per kilogram since June 2015 in most markets across the country. The wholesale price index (WPI) for pulses has also risen year over year—for example, by about 22.8 percent in May and 33.7 percent in June of 2015, with even steeper increases for pigeon pea of 30 percent and 42 percent for the same two months. This 2014–2015 rise in wholesale prices has been primarily due to a nearly 10 percent fall in the output of pulses over the same year.

Looking at the recent 2014–2015 period, production of chickpea and pigeon pea dropped steeply again, by about 23 percent and 15 percent, respectively. Consequently, prices spiraled upward, despite a 27 percent increase in the importation of pulses in 2014–2015, which reached a high of 4.6 million tons (Lingareddy 2015). Theoretically, the persistently rising prices should have triggered a positive supply response in pulses, but a possible lack of price transmission from market to farmgate could explain why the supply response from farmers has been muted (see Rahman 2015). This is discussed further in Chapter 5.

The overall results of studies of price elasticity seem to show that pulse growers do not respond to commercial incentives (Tuteja 2006). The nonprice factors that Tuteja (2006) highlights include the previous year's acreage and yield, availability of improved seeds and irrigation, rainfall, resistance of crop to pest attacks, extension services, home consumption, availability of alternate crops, credit, and assured market. Several studies show that nonprice factors are comparatively important as determinants of acreage and supply response (Chopra and Swamy 1975; Chopra 1982; Deshpande and Chandrashekar 1982; Acharya 1988; Sadasivam 1993; Dhindsa and Sharma 1997). In Tuteja (2006) nonprice factors dominated over price factors in determining acreage allocated to pigeon pea, both in major growing states and at the all-India level. The results of the acreage response model for other *kharif* pulses, such as green gram and black matpe, were similar. Importantly, the elasticities of relative price were found to be low and insignificant. In contrast, the amount of presowing rainfall has been shown to have an impact on the area planted in *kharif* pulses, again both in the majority of the referred states and at the all-India level. Relative yield also had an impact on the area allocated to black matpe in Madhya Pradesh, Uttar Pradesh, and at the all-India level. The nature of supply response to prices (ostensibly not large enough) raises questions about how increases in minimum support price (MSP) in pulses would result in increases in the supply of pulses. A supply response would likely be forthcoming only if MSP were to be raised by a large magnitude, which for various reasons might not be feasible (for example, fiscal capacity, administration).

The Link between Production and Irrigation

According to Lingareddy (2015), growth in pulse production has historically been sluggish, with the exceptions of growth during the 1950s and 2000s. Production has been highly volatile from year to year, mainly due to wide fluctuations in yields. Reddy (2009b) indicates that the cultivation of pulses on rainfed and marginal lands has contributed to their low and uncertain yields. Recent statistics continue to bear this out. In the triennium ending 2011–2012, for example, only about 15 percent of land planted in pulses was irrigated, as compared with more than 90 percent of land planted in wheat and close to 60 percent of land planted in rice. Among pulses, during the same period the irrigated area of chickpea was 34 percent, while for pigeon pea it was only 4 percent (India, Ministry of Agriculture 2015). Sadasivam (1989) showed that the substitution effect in favor of cereals was more pronounced in the *rabi* season in states that had assured water supply, such as Bihar, Haryana, Punjab, Uttar Pradesh, and West Bengal—essentially the wheat belt of India. As discussed earlier, the improvement in area under cultivation during the late 1970s and 1980s was mainly on account of a shift in pulses cultivation to the drylands of India's central and southern regions (Sadasivam 1989).

We have seen that intensive irrigation leads farmers to switch from pulses to other crops. However, provision of protective irrigation can possibly be quite effective for increasing pulse production. The protective irrigation systems are designed and operate on the principle that the available water has to be spread thinly over a large area and in an equitable manner. The objective is to reach as many farmers as possible and to protect against crop failure and famine. The amount of water a farmer would receive under protective irrigation would be insufficient to cover full crop water requirements on all of his land for an average rainfall year.

The prime minister of India recently launched a big program called Har Khet Ko Paani (Water to Every Farm) initiative. Under the Pradhan Mantri Krishi Sinchayee Yojana (PMKSY, the Prime Minister Agriculture Irrigation Plan), it can be helpful to accord priority to the provision of lifesaving irrigation in pulse-growing areas. A similar provision in the central tribal region can help bring a part of the 11 million hectares of rice fallows under pulses. Access to even one or two lifesaving irrigations over the life of the crop can give a quantum boost to pulse production and productivity, and significantly reduce production risks. Thus, investment in lifesaving irrigation in pulse-growing and rice fallow areas of India can be an important choice. The uptake of irrigation conditional on growing pulses would depend on the provisions built in irrigation programs. Otherwise, farmers might migrate to other crops if irrigation were to become available.

Ineffectiveness of Government Pricing Policies

On the pricing policy front, the government has regularly hiked its minimum support prices for pulses, but it has not been able to conduct active procurement operations. Without procurement, the price support system for pulse farmers has been ineffective. The National Agricultural Cooperative Marketing Federation of India (NAFED), the agency entrusted with the limited procurement of pulses, has not been very active (Lingareddy 2015).[3] Grover and Singh (2012) indicate that in the absence of any procurement support, the acreage under pulses cannot be augmented significantly. However, the implementability and effectiveness of a large procurement program in pulses cannot be taken for granted. Below we explain why MSP without large-scale procurement may be counterproductive, and yet the introduction of a major procurement program in pulses is fraught with major challenges.

The government of India has indeed tried to incentivize an increase in pulse production and productivity by raising its MSP. On a couple of occasions, the MSP was increased very substantially—for example, it has been increased by more than 50 percent since 2010. For the 2015–2016 crop year (July–June), the agriculture ministry announced up to a 6 percent increase in MSP, including a bonus of 200 rupees per quintal. With the increase, the MSP of black matpe touched 4,625 rupees per quintal for 2015–2016, compared with 4,350 rupees per quintal the previous year. Still, a commensurate supply response to such increases in the MSP has not been observed.

Based on the recommendations of the Commission for Agricultural Costs and Prices (CACP), India's Department of Agriculture and Cooperation declares MSPs for 22 crops before their sowing seasons each year. The MSP is aimed at giving farmers a guaranteed price and an assured market to protect them from price fluctuations. This is expected to encourage higher investment and adoption of modern farming practices. MSPs for rice and wheat were started with the introduction of high-yielding varieties, amid fears that a glut on the market would adversely affect farmers. These two commodities are now in surplus, and MSPs are also set for deficit crops like pulses.

---

3 NAFED is the public procurement agency for pulses and oilseeds, similar to the Food Corporation of India (FCI).

With MSPs announced based on the recommendations of CACP, we argue that it makes a difference whether the crop is in surplus (supply greater than demand at MSP) or is in deficit (demand greater than supply at the announced MSP). For pulses, the demand is usually greater than the supply at the announced MSP—that is, there is a deficit (shown in Figure 3.3). There are three possible cases of supply: (1) perfectly inelastic supply (line CD in the figure); (2) elastic supply (shown as ST in the figure); and (3) piecewise elastic supply (shown as CFT). The piecewise inelasticity in supply can come from several factors, such as lack of substitutes in production or lack of inputs. Depending on the season and area, there are competing crops for pulses—for example, soybean in Maharashtra, wheat in several states, cotton in some states, and some other commercial crops such as chilies. In the case of pulses, the channel that we believe is salient is the riskiness where unless price rise covers for risk premium, the supply response may not follow. Only when the magnitude of price rise is substantive, can one expect a supply response.

In the market, trade takes place between farmers and traders at or around the MSP, with or without procurement by the government. The easiest way to understand the situation of a deficit crop is by considering a perfectly inelastic supply. Referring to Figure 3.3, if MSP = M1, farmers receive much less than the potential price given by the demand curve equal to M3. Under all MSPs up to M3, for the quantity given by SC, the farmer should be receiving prices higher than the MSP. For example, at an MSP of M1, the pulse farmer gets less per unit by the amount represented by line PG. Above M3, the crop will be not a deficit commodity but a surplus one. If the MSP is announced before sowing and brings in a supply response, the curve will shift to the right, but farmers will still get a lower price than without the MSP unless the MSP is raised significantly, to the level of M3 or a corresponding level in relation to the new supply curve.

Even if the supply curve is inelastic domestically, imports could compensate. In this case, there are two possibilities: (1) global markets can bridge the deficit, or (2) global markets are thin and cannot meet the requirements. If the deficit is bridged with imports, there may not be a gap between the MSP and the potential price. If global markets are insufficient, the wedge between the actual and potential farmgate price will be sustained. With a piecewise elastic supply or an inelastic supply, an MSP higher than M1 and lower than M2 or M3 (depending on the case) should not transmit to market prices. It will do so only if traders' margins are unchecked. With an elastic supply (ST in Figure 3.3), a higher MSP can affect the level of excess demand more than in the inelastic case and also can affect market price.

FIGURE 3.3 Supply, demand, minimum support price, and farmgate prices in case of deficit crops such as pulses

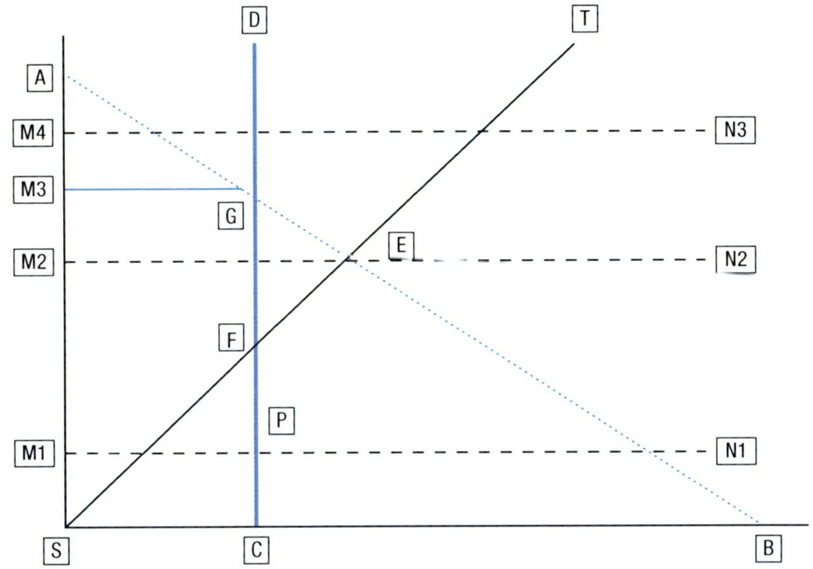

Source: Authors.

Until now we have assumed that the seller's price is at the MSP. Given the deficit at MSP, should not the price at which farmers sell to traders rise? Given the nature of the pulses market, we argue that it does not because MSP works as a focal point of tacit collusion among traders, who offer farmers a price that is near the MSP (Figure 3.4 plotted for a wedge between farmgate prices and the MSP). CACP data show that farmgate prices for pulses are heavily centered around the MSP even though there is limited or no procurement. This is true for all pulses and is positively skewed for prices faced by comparatively large farmers, due to their greater bargaining power (an issue that is discussed further in Chapter 5). In this situation, increasing the MSP would raise the farmer's price, and without procurement, the fiscal costs would be nil. Moreover, this channel is independent of what the market price is. If there is pass-through to the consumer price, the government could mitigate the price rise by holding credible stocks to calm the markets. It is also likely that the market (retail) price is determined by supply and demand and is not a function of the farmgate prices in the same period, given short-run inelasticity. In addition, many times farmers find out the MSP after sowing, which also leads to inelasticity.

Note that in the recent past, very few years have witnessed average farmgate prices going below the MSP. Since 2000, the farm harvest price (FHP) of chickpeas has been around the MSP or just a little lower only once: in 2013–2014. For pigeon peas the two prices have been similar only three times in the past 17 years: the FHP was marginally below the MSP in 2011–2012 and 2012–2013 and almost the same in 2004–2005. These are the years with significant increases in the MSP as well as spikes in imports. Hence, it is not unreasonable to suggest that without the announcement of MSPs for pulses, the farmgate prices can effectively rise because of the tacit collusion among traders (see Rahman 2015) at MSP might get broken.

Price supports therefore work differently for pulses than for rice, wheat, or oilseeds:

- Unlike that of rice and wheat, pulse production is less than the annual demand and there is no or very limited procurement at the MSP. Furthermore, unlike oilseeds, there is not much availability of pulses in the international markets either, certainly not at lower prices.
- Even when the MSP for pulses has been raised significantly, it has stayed below the market price of pulses in every single year since 2000.

We contend that when the support price of pulses is near or below the market price and the opportunities to import them cheaply from other countries are limited, the MSP helps traders more than producers. It acts as a focal point, or a Schelling point, for pulse traders to facilitate implicit collusion at prices below what the market price otherwise would be. There is clear clustering of farmgate prices around the MSP (Figure 3.4) that is unlikely without this sort of tacit collusion. It is possible that farmers may receive higher prices if the MSP were not announced and there were no anchors for traders to collude around.

Several studies already mentioned, such as Tuteja (2006), show supply of pulses to be price irresponsive (at least in the ranges in which they have moved historically). Small or moderate increases in MSP are not likely to bring forth significant supply response in pulses. Combining with issues relating to the extent MSP will have coverage will be timely given the experience from rice and wheat. There is a sheer lack of physical and institutional infrastructure to implement procurement-based support prices for pulses mandates research before a policy stance on this becomes clear.

Hence, several different tasks regarding pricing in pulses remain to be tackled going forward. Apart from research to assess the feasibility and

FIGURE 3.4 Distribution of farmgate prices

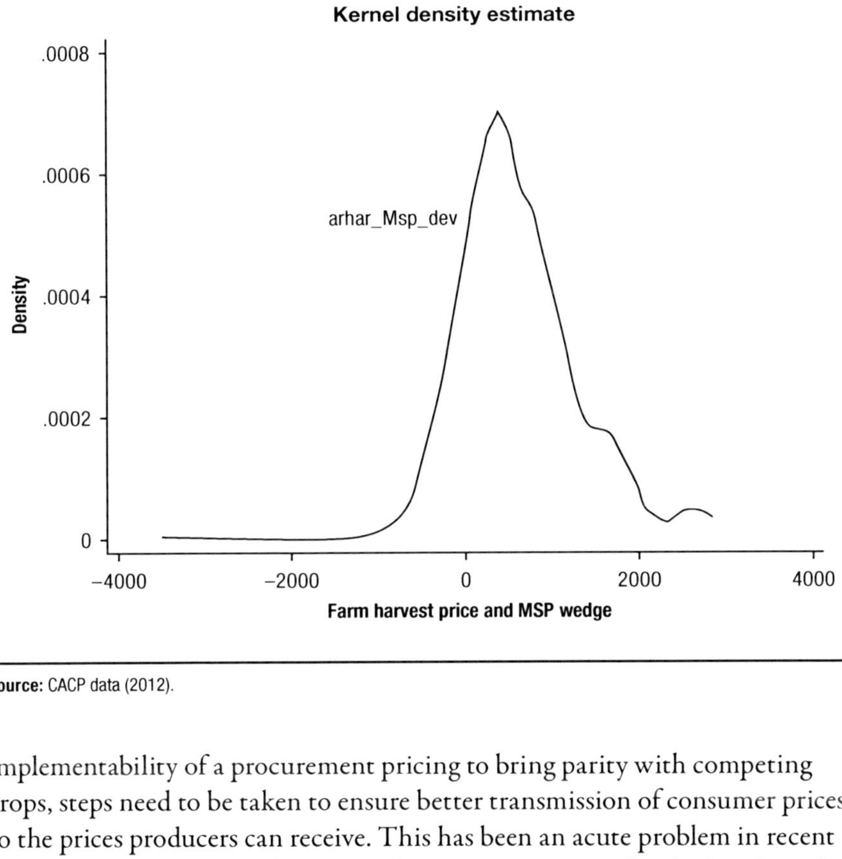

Source: CACP data (2012).

implementability of a procurement pricing to bring parity with competing crops, steps need to be taken to ensure better transmission of consumer prices to the prices producers can receive. This has been an acute problem in recent episodes of pulse price spikes, as producer prices continued to be benchmarked to the MSP while the retail prices increased many times. Fixing this problem would require establishing direct farm-to-fork or firm-farm links. The role of farmer producer organizations, which balance the bargaining power of farmers, can be instrumental in this regard. The role of processing too, with its backward links, is very important.

Beyond Support Prices: Paying Pulse Growers and Pulse-Growing Areas for Ecosystem Services

Among protein-rich foods, pulses have the lowest carbon and water footprint. In addition, pulses improve soil health by naturally fixing atmospheric nitrogen in the soil; growing pulses reduces the need for application of nitrogenous fertilizer, especially urea, in the subsequent crop. Thus pulses provide valuable environmental services (Dudeja and Duhan 2005). Thanks to India's diverse agroclimatic conditions, pulses are grown in the country throughout the year. Several benefits from pulses are particularly important, such as their role in crop rotation and in intercropping, because they help improve soil fertility by reducing soil pathogens and fixing nitrogen. Studies show that because of these factors, the yield of a crop that follows pulses can increase by up to 20–40 percent (Pande and Joshi 1995).

Changes in soil fertility have been assessed for different crops—for example, maize (Dwivedi et al. 2015; Kumar et al. 2015). Lower usage of fertilizer, pesticide, and irrigation further makes pulses an environmentally sustainable crop group. Saddled with a huge fertilizer subsidy burden and food safety issues from excessive chemical use in farming, India can benefit greatly from these roles of pulses. Assessing the value of environmental services provided by pulses and devising mechanisms to reward farmers or pulse-growing areas for these ecosystem services could be one policy option. Paying individual farmers may be logistically difficult, but it could be feasible to pay pulse-growing areas by offering them additional resources for investment in agriculture, irrigation, or extension in the same way that the fourteenth finance commission of India has offered states incentives to maintain and increase area under the forests.

Lessons from Recent Increases in Pulse Supply

After studying the recent increases in supply of some pulses, Reddy, Bantilan, and Mohan (2013) suggest lessons that can be drawn for other crops. They argue that even though pulse production has increased significantly during the past decade, continuing fast growth will be a bigger challenge for researchers, extension agencies, and policy makers. For some crops, such as oilseeds, earlier experience shows that most of the success is short-lived if production technology is not aligned with policy support (Reddy 2009a). Reddy, Bantilan, and Mohan (2013) examined the factors behind the fast growth in the production of pigeon pea nationally and the production of chickpea in Andhra Pradesh, which they regard as examples of success. They cite the introduction of the chickpea crop into nontraditional areas like the southern Indian states as an example of a technological and institutional breakthrough to be replicated in other pulses. They highlight a set of opportunities, as well as programs and practices, that led to the successful growth in Andhra Pradesh, including the introduction of chickpea into black cotton soils, the availability of plenty of *rabi* fallow lands, the adoption of short-duration and high-yielding varieties like KAK-2 and JG-11, and the well-developed land-lease market that enabled large-scale mechanization to cope with labor shortages in villages.

At a program level, the success of chickpea in Andhra Pradesh highlighted the importance of (1) successful government programs like the National Food Security Mission (NFSM) in increasing pulse production; (2) the development and distribution of improved seed through semiformal seed systems and farmers' participatory varietal selection (FPVS); (3) emphasis on abiotic and biotic stress management to increase stability in area and yields through integrated approach; (4) increased availability of subsidized improved seed, micronutrients like sulfur and gypsum, and the popularization of herbicides and farm machinery to cope with labor shortages; and, last, (5) the development of market information systems and warehouse infrastructure, including state-of-the-art postharvest management and cold storage, to enhance credit availability and establish markets. In Andhra Pradesh the yield of chickpea increased between 1987 and 2008 from 393 kilograms (per hectare) to 1,375 kilograms, while the area cultivated increased from 52,200 hectares to 542,000 hectares, which resulted in production increase from 19,900 tons to 730,700 tons during the same period.

This finding is important in light of the fact that for pulses there have been relatively few significant technological breakthroughs until now due to peculiar problems like indeterminate plant type, low response to fertilizers, and management practices. However, after experiencing a steep rise in prices and declining per capita availability of pulses, governments have encouraged pulse production through various programs, including the Integrated Scheme of Oilseeds, Pulses, Oil Palm, and Maize (ISOPOM) and the National Food Security Mission. These government efforts have been supported by various research bodies, such as the national agricultural research systems (NARS) and the International Crops Research Institute for the Semi-Arid Tropics (ICRISAT). All of this has resulted in some improvement in the production of major pulses, including chickpea and pigeon pea, although this growth has only been significant since 2001. The growth rate of chickpea is currently 6.32 percent per year and for pigeon pea it is 2.05 percent, while the total growth in production of (all) pulses is 3.35 percent—a rate that is well ahead of the population growth but way below the growth in demand.

Conclusion

This study of the supply dynamics of pulses reveals that the traditional areas for pulse production have been switching to other crops as pulses have moved to nontraditional areas. This is reflected in the movement of pulses from northern to southern and from eastern to western zones, with central India becoming the hub for pulse farming. Andhra Pradesh, Karnataka, Madhya Pradesh, Maharashtra, and Rajasthan are emerging as the most promising states in pulse production. Madhya Pradesh dominates in chickpea, and Maharashtra dominates in pigeon pea. Identifying the factors behind this transition is an important area for future research. Different factors have been proposed, such as the spread of some new varieties and a significant increase in support prices for pulses. For example, the observed patterns point to the likelihood that technology has played a crucial role in the shift in areas allocated to pulses from traditional into nontraditional areas.

The states that do not have any highly commercialized or highly profitable crops seem to be the ones where the adoption of pulses has been easiest. For instance, the absence of any profitable star crop in the Telangana and Rayalaseema regions of Andhra Pradesh is facilitating chickpea cultivation, so much so that they have emerged as the major producer of chickpea in India. In the same way, low-productive cotton in northern Karnataka and barley in Madhya Pradesh are being replaced by chickpea. Reddy (2009b) presents the case of chickpea's rapid expansion in Andhra Pradesh as a suitable example to emulate for production growth in other pulse crops.

Dividing the five decades into four periods, we see substantial temporal variation among the periods in the land area allocated to and production of pulses. For pigeon pea, the area, production, and yield all increased during the initial and matured phases of the Green Revolution, but they declined quite significantly during the postliberalization period. It was just the opposite for chickpea, for which the post–trade spike period (2000–2010) showed an increase in area, production, and yield. During this most recent period, net yield increased for all pulses, for the first time by a double-digit percentage.

Both technical and environmental considerations play a role in the movement of pulse farming. The introduction of short-duration varieties of chickpea, for example, has contributed to the increase in its area and production. The July rainfall has been shown by different studies to be negatively associated with area allocation to pigeon pea, implying that proper soil and water conservation measures need to be taken to ensure that flooding does not occur. At the same time, water logging could certainly play a role in diminishing area planted, especially in deep black soil.

Trends of regional specialization and geographic continuity are emerging in regard to the area and production of pulses, which are visible at the regional and state level. The common features observed across the currently leading pulse-producing states are rainfed conditions, absence of irrigation, and general lack of alternative profitable crops. Based on the discussion in this chapter, the following are some potential approaches for increasing pulse production:

- The pulses being cultivated predominantly in the marginal and rainfed regions under resource-starved conditions require a different approach to increase their area, production, and productivity. Indeed, research efforts have already shifted to develop varieties that can address the challenges of the nontraditional areas.
- Apart from concentration in the marginal environments, the diversity in determinants of production and consumption across space also needs to be internalized in policy. One implication of these spatial dynamics is that research needs to shift its relative resource allocation from the northern zone to the southern and central zones, particularly for chickpea and pigeon pea but also for pulses in general. This is elaborated in Chapter 4 on technology.
- In addition, research must be undertaken to find the potential of pulse production in the rainfed rice fallow systems spread across the states of Bihar, Chhattisgarh, Jharkhand, Madhya Pradesh, and Odisha. Post–rainy season fallows might comprise both the biggest challenge and the greatest opportunity for increasing pulse production. Alternative cropping patterns and the adoption of various technologies, coupled with seed availability, could increase pulse production in these marginal areas.
- Also, there is need for further research to assess the feasibility and risks of policies increasing the MSP and expanding procurement in a deficit crop such as pulses.

References

Acharya, S. S. 1988. *Agricultural Production, Marketing, and Price Policy in India: A Study of Pulses*. New Delhi: Mittal Publications.

CACP (Commission on Agricultural Costs and Prices). 2012. Reports on Price Policy. Accessed June 2016. Available at http://cacp.dacnet.nic.in/KeyBullets.aspx?pid=39.

Chand, R. 2008. "The Global Food Crisis: Causes, Severity, and Outlook." *Economic and Political Weekly* 43 (26): 137–144.

Chopra, K. 1982. "Pulse Production in India: A State Wise Analysis." *Indian Journal of Agricultural Economics* 37 (3): 371–380.

Chopra, K., and G. Swamy. 1975. *Pulses: An Analysis of Demand and Supply in India, 1951–1971.* New Delhi: Sterling Publishers.

Deshpande, R. S., and H. Chandrashekar. 1982. "Growth and Supply Response of Slow Growth Crops—A Case of Pulses." *Indian Journal of Agricultural Economics* 37 (3): 386–393.

Dhindsa, K. S., and A. Sharma. 1997. "A Regional Analysis of Growth and Supply Response of Pulses—A Study of Punjab." *Indian Journal of Agricultural Economics* 52 (1): 87–100.

Dudeja, S. S., and J. S. Duhan. 2005. "Biological Nitrogen Fixation Research in Pulses with Special Reference to Mungbean and Urdbean." *Indian Journal of Pulses Research* 18 (2): 107–118.

Dwivedi, A., I. Dev, V. Kumar, R. S. Yadav, and M. Yadav. 2015. "Potential Role of Maize-Legume Intercropping Systems to Improve Soil Fertility Status under Smallholder Farming Systems for Sustainable Agriculture in India." *International Journal of Life Sciences Biotechnology and Pharma Research* 4 (3): 145–157.

Grover, D. K., and J. M. Singh. 2012. "Possibilities and Constraints for Increasing the Production of Pulses in Punjab and Impact of National Food Security Mission on Pulses." Research report to the Directorate of Economics and Statistics. India, Ministry of Agriculture. New Delhi.

IIPR (Indian Institute of Pulses Research). 2011. *IIPR Vision 2030.* Kanpur, India: Kanpur Indian Institute of Pulses Research (ICAR).

India, Ministry of Agriculture. 2015. *Agricultural Statistics at a Glance 2014.* Department of Agriculture and Cooperation. Directorate of Economics and Statistics. New Delhi: Oxford University Press.

———. 2014. *Agricultural Statistics at a Glance.* Department of Agriculture and Cooperation. Directorate of Economics and Statistics. New Delhi: Oxford University Press.

India, Ministry of Agriculture and Cooperation. 2010. *Agricultural Statistics at a Glance.* Directorate of Economics and Statistics. Accessed January–February 2016. http://eands.dacnet.nic.in/Advance_Estimate-2010.htm.

Joshi, P. K., M. Asokan, and M. C. S. Bantilan. 1999. "Silent Chickpea Revolution in Non-traditional Areas—Some Evidences from Andhra Pradesh." *Indian Journal of Agricultural Economics* 54 (4): 533–544.

———. 2005. "Chickpea in Non-traditional Area: Evidence from Andhra Pradesh." In *Impact of Agricultural Research: Post-Green Revolution Evidence from India*, edited by M. C. S. Bantilan, 115–129. New Delhi: National Centre for Agricultural Economics and Policy Research; Patancheru, India: International Crops Research Institute for the Semi-Arid Tropics.

Joshi, P. K., and R. Saxena. 2002. "A Profile of Pulses Production in India: Facts, Trends, and Opportunities." *Indian Journal of Agricultural Economics* 57 (3): 326–339.

Kaushal, N., and F. M. Muchomba. 2015. "How Consumer Price Subsidies Affect Nutrition." *World Development* 74: 25–42.

Kishore, A., and S. Chakrabarti. 2015. "Is More Inclusive More Effective? The 'New Style' Public Distribution System in India." *Food Policy* 55: 117–130.

Kumar, R., A. N. Paslawar, U. Singh, and R. K. Nagar. 2015. "Effect of Conservation Tillage on Growth, Production, and Energetics of Pigeonpea-Based Intercropping System under Rainfed Area." *The Ecoscan* 9 (1–2): 153–157.

Lingareddy, T. 2015. "Pulses: Need for Production Expansion." *Economic and Political Weekly* 50 (39), August 29.

Materne, M., and A. A. Reddy. 2007. "Commercial Cultivation and Profitability." In *The Lentil: An Ancient Crop for Modern Times*, edited by S. S. Yadav, D. L. Mc Neil, and P. C. Stevenson, 73–186. Rotterdam, Netherlands: Springer.

Pande, S. P., and P. K. Joshi. 1995. *Constraints and Prospects of Legumes in Northern India and the Terai Region of Nepal*. Patancheru, India: International Crops Research Institute for the Semi-Arid Tropics.

Rahman, A. 2015. "Is Price Transmission in the Indian Pulses Market Asymmetric?" *Journal of Quantitative. Economics* 13: 129–146.

Reddy, A. A. 2006. "Growth and Instability in Chickpea Production in India: A State Level Analysis." *Agricultural Situation in India* 230–245.

———. 2009a. "Policy Options for India's Edible Oil Complex." *Economic and Political Weekly* 44 (4): 22–24.

———. 2009b. "Pulses Production Technology: Status and Way Forward." *Economic and Political Weekly* 44 (52): 73–80.

Reddy, A. A., M. C. S. Bantilan, and G. Mohan. 2013. "Pulses Production Scenario: Policy and Technological Options." ICRISAT Policy Brief 26. International Crops Research Institute for the Semi-Arid Tropics.

Sadasivam, S. 1989. "Pattern of Pulses Production—An Analysis of Growth Trends." *Economic and Political Weekly* 24 (51–52): December 23.

———. 1993. "Price and Non-Price Factors in Input Use and Output Supply—A Case of Gram in Uttar Pradesh." *Indian Economic Review* 28 (1, January–June): 73–83.

Sadavatti, P. 2007. "An Econometric Analysis of Demand and Supply Response of Pulses in India." *Karnataka Journal of Agricultural Sciences* 20 (3): 545–550.

Shiyani, R. L., P. K. Joshi, M. Asokan, and M. C. S. Bantilan. 2000. "Adoption of Improved Chickpea Varieties: Evidences from Tribal Region of Gujarat." *Indian Journal of Agricultural Economics* 55 (2): 159–171.

Singh, L., Y. S. Chauhan, C. Johansen, and S. P. Singh, eds. 1996. *Prospects for Growing Extra-Short-Duration Pigeonpea in Rotation with Winter Crops*. Proceedings of the IARI (Indian Agricultural Research Institute)/ICRISAT (International Crops Research Institute for the Semi-Arid Tropics) workshop and monitoring tour, October 16–18, 1995. New Delhi.

Srivastava, S. K., N. Sivaramane, and V. C. Mathur. 2010. "Diagnosis of Pulses Performance of India." *Agricultural Economics Research Review* 23: 137–148.

Suhasini, P., V. R. Kiresur, G. D. N. Rao, and M. C. S. Bantilan. 2009. *Adoption of Chickpea Cultivars in Andhra Pradesh: Pattern, Trends, and Constraints*. Patancheru, India: International Crops Research Institute for the Semi-Arid Tropics (ICRISAT).

Tuteja, U. 2006. "Growth Performance and Acreage Response of Pulse Crops: A State Level Analysis." *Indian Journal of Agricultural Economics* 61 (2): 219–237.

Appendix

TABLE 3A.1 Statewise competing crops of chickpea and pigeon pea

	Competing crops of	
State	Chickpea	Pigeon pea
Andhra Pradesh	*Rabi* sorghum	Paddy, cotton, pearl millet, groundnut, maize
Assam	Wheat	Paddy, maize
Bihar	Wheat	Paddy, maize
Gujarat	Wheat	Paddy, cotton, pearl millet, groundnut, maize
Haryana	Wheat	Paddy, cotton, pearl millet
Himachal Pradesh	Wheat	Paddy, maize
Jammu and Kashmir	Wheat	Paddy, maize
Karnataka	*Rabi* sorghum	Paddy, cotton, maize
Madhya Pradesh	*Rabi* sorghum, wheat	Paddy, cotton, pearl millet, groundnut, maize
Maharashtra	*Rabi* sorghum, wheat	Paddy, cotton, pearl millet, groundnut, maize, soybean
Odisha	Wheat	Paddy, cotton, groundnut, maize
Punjab	Wheat	Paddy, cotton, maize
Rajasthan	Mustard	Paddy, cotton, pearl millet, groundnut, maize
Tamil Nadu	*Rabi* sorghum	Paddy, cotton, groundnut, maize
Uttar Pradesh	Wheat	Paddy, pearl millet, maize
West Bengal	Wheat	Paddy, groundnut, maize

Source: Authors' assessment.

TABLE 3A.2 Decadal change in chickpea production, 1960–2010 (%)

State	Pre– and initial phase of Green Revolution (1960–1970)	Matured phase of Green Revolution (1971–1990)	Postliberalization period (1991–2000)	Post–trade spike period (2001–2010)
Northern Zone				
Haryana	−13.6	−54.4	−51.6	−63.2
Himachal Pradesh	−60.7	−70.0	−22.7	−74.0
Punjab	−52.3	−87.1	−80.3	−65.0
Uttar Pradesh	4.9	−35.5	−28.2	−37.0
Southern Zone				
Andhra Pradesh	−33.3	35.3	288.0	729.9
Karnataka	124.0	−14.5	105.9	172.3
Tamil Nadu	50.0	86.6	35.7	−11.1
Eastern Zone				
Assam	—	30.0	−23.0	30.0
Bihar	−27.6	−27.3	−70.8	−40.5
Odisha	156.2	109.0	−37.4	57.2
West Bengal	19.3	−76.4	−32.4	23.1
Western Zone				
Gujarat	−60.2	28.3	94.0	92.4
Maharashtra	−17.1	186.5	49.5	106.7
Rajasthan	−3.5	−14.0	123.8	−55.3
Central Zone				
Madhya Pradesh	1.1	74.6	66.6	5.1
All India	−9.7	−18.0	39.1	12.3

Source: Authors' calculations.

Note: — = data not available.

TABLE 3A.3 Decadal change in pigeon pea production (%)

States	Pre– and initial phase of Green Revolution (1960–1970)	Matured phase of Green Revolution (1971–1990)	Postliberalization period (1991–2000)	Post–trade spike period (2001–2010)
Northern Zone				
Haryana	—	853.8	–19.4	–7.9
Punjab	—	—	–63.1	27.8
Uttar Pradesh	–0.3	–7.5	–18.3	–46.2
Southern Zone				
Andhra Pradesh	62.6	–24.9	97.8	85.1
Karnataka	23.2	53.5	11.8	77.5
Tamil Nadu	–16.2	416.1	–48.2	–65.5
Eastern Zone				
Assam	60.0	112.5	–11.1	–19.2
Bihar	37.5	–44.0	5.9	–58.4
Odisha	181.8	437.1	–25.9	39.5
West Bengal	62.5	–78.2	–63.5	67.7
Western Zone				
Gujarat	16.9	419.8	36.5	–14.1
Maharashtra	–3.4	92.6	10.2	28.3
Rajasthan	38.8	80.0	88.6	–55.9
Central Zone				
Madhya Pradesh	21.1	44.6	–37.6	–10.3
All India	10.6	42.6	–5.8	7.6

Source: Authors' calculations based on data from India, Ministry of Agriculture 1950–2010.

Note: — = data not available.

TABLE 3A.4 Grouping of states of India for area allocation and production of total pulses

	Total pulses	
Typology of states	Area	Production
Gaining ground	Andhra Pradesh Karnataka Madhya Pradesh Maharashtra	Andhra Pradesh Karnataka Madhya Pradesh Maharashtra
Losing ground	Assam Bihar Gujarat Himachal Pradesh Haryana Punjab West Bengal Uttar Pradesh Tamil Nadu Odisha	Assam Bihar Gujarat Himachal Pradesh Haryana Punjab West Bengal Uttar Pradesh Himachal Pradesh Tamil Nadu Odisha
Status quo states	—	—

Source: Authors' calculations based on data from India, Ministry of Agriculture, 1950–2010.

Note: — = data not available.

TABLE 3A.5 Pulses' total production, yield, and area under production (1960–2010)

	Total area (in thousands of hectares)	Total production (in thousands of metric tons)	Total yield (in kilograms per hectare)
1960–1961	23,563	12,704	539
1961–1962	24,243	11,755	485
1962–1963	24,265	11,528	475
1963–1964	24,186	10,073	416
1964–1965	23,875	12,417	520
1965–1966	22,717	9,944	438
1966–1967	22,121	8,347	377
1967–1968	22,649	12,102	534
1968–1969	21,264	10,418	490
1969–1970	22,023	11,691	531
1970–1971	22,534	11,818	524
1971–1972	22,151	11,094	501
1972–1973	20,915	9,907	474
1973–1974	23,427	10,008	427
1974–1975	22,024	10,014	455
1975–1976	24,454	13,040	533
1976–1977	22,983	11,361	494
1977–1978	23,497	11,973	510
1978–1979	23,657	12,183	515
1979–1980	22,259	8,572	385
1980–1981	22,457	10,627	473
1981–1982	23,843	11,507	483
1982–1983	22,833	11,857	519
1983–1984	23,542	12,893	548
1984–1985	22,737	11,963	526
1985–1986	24,418	13,361	547
1986–1987	23,156	11,707	506
1987–1988	21,559	11,040	512
1988–1989	23,146	13,849	598
1989–1990	23,415	12,858	549
1990–1991	24,662	14,265	578
1991–1992	22,543	12,015	533

	Total area (in thousands of hectares)	Total production (in thousands of metric tons)	Total yield (in kilograms per hectare)
1992–1993	22,360	12,815	573
1993–1994	22,250	13,305	598
1994–1995	23,028	14,038	610
1995–1996	22,283	12,310	552
1996–1997	22,447	14,244	635
1997–1998	22,871	12,979	567
1998–1999	23,501	12,162	518
1999–2000	21,116	13,418	635
2000–2001	20,348	11,076	544
2001–2002	22,008	13,368	607
2002–2003	20,496	11,125	543
2003–2004	23,458	14,905	635
2004–2005	22,763	13,130	577
2005–2006	22,391	13,384	598
2006–2007	23,192	14,198	612
2007–2008	23,633	14,762	625
2008–2009	22,094	14,566	659
2009–2010	23,282	14,662	630

Source: Authors' calculations based on data from India, Directorate of Economics and Statistics, Ministry of Agriculture.

FIGURE 3A.1 Pigeon pea districts, 1966–2010

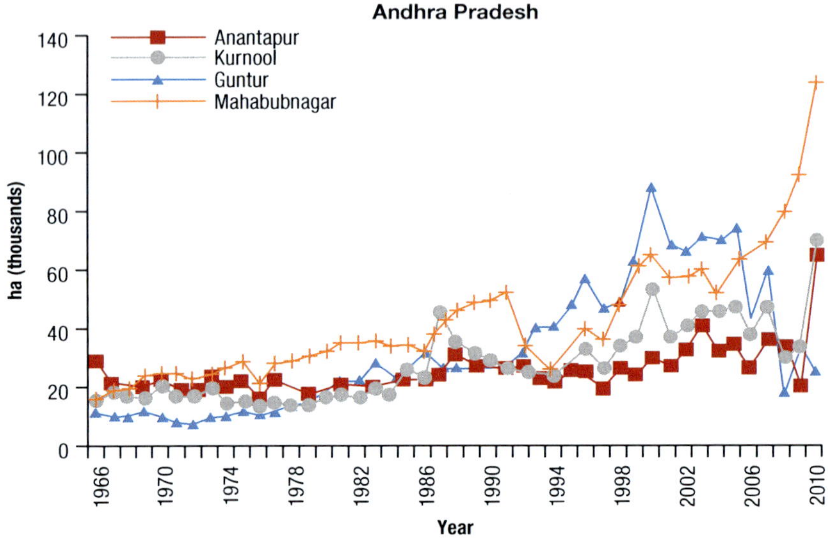

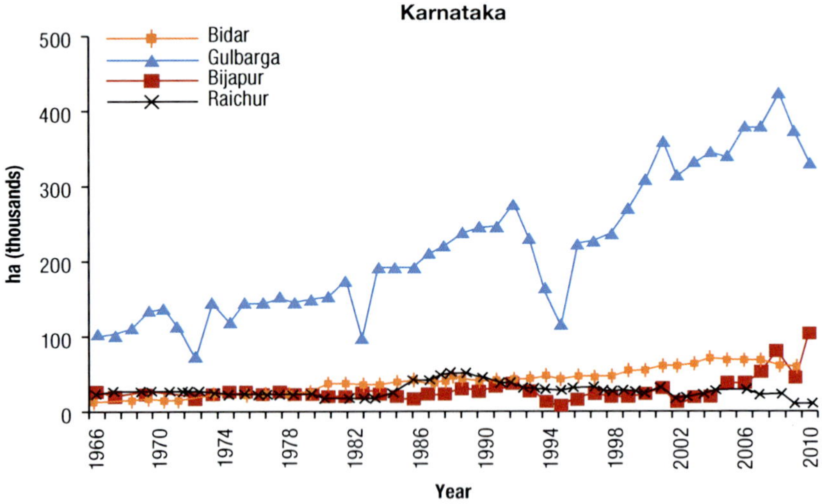

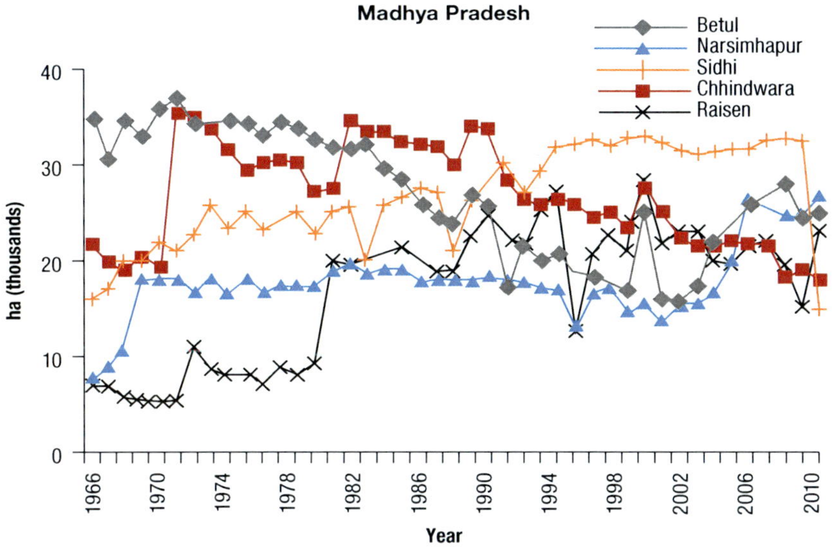

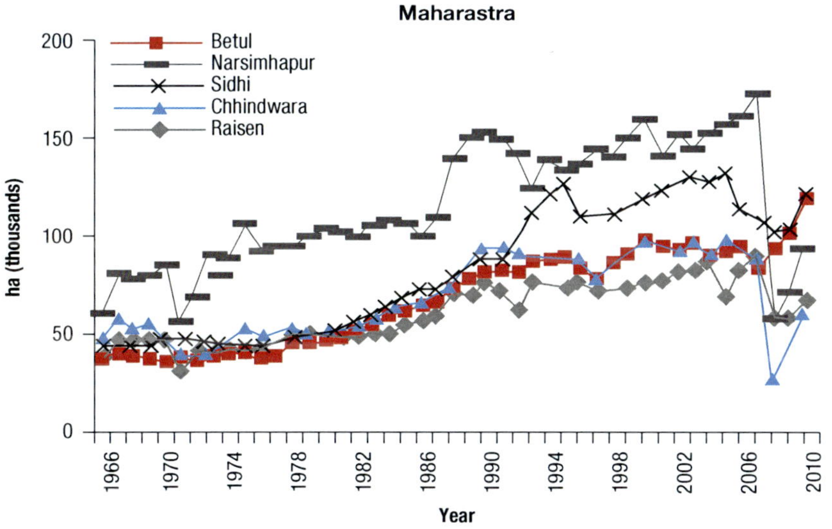

Source: Authors' calculations based on data from India, Directorate of Economics and Statistics, Ministry of Agriculture.

FIGURE 3A.2 Chickpea districts, 1966–2010

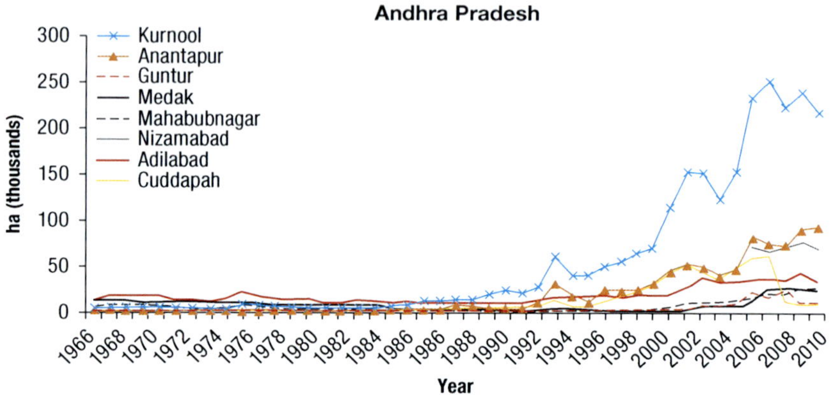

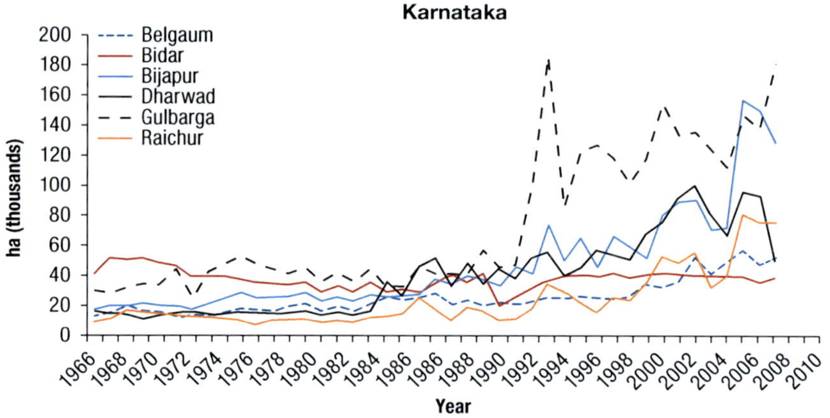

Madhya Pradesh

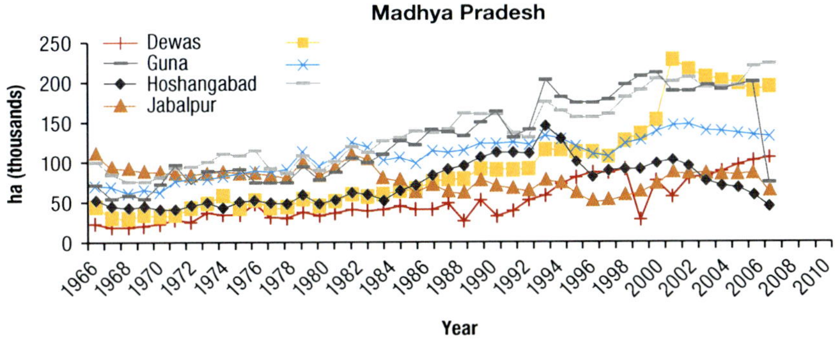

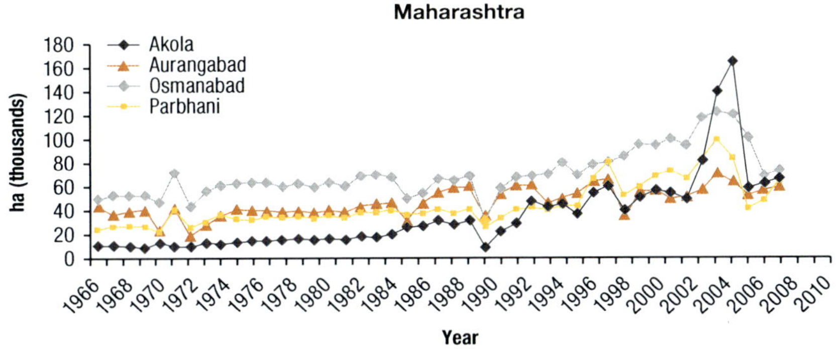

Source: Authors' calculations based on data from India, Directorate of Economics and Statistics, Ministry of Agriculture.

FIGURE 3A.3 Shift in chickpea area and production across states, 1970 and 2007

Chickpea area ('000 Ha) in 1970

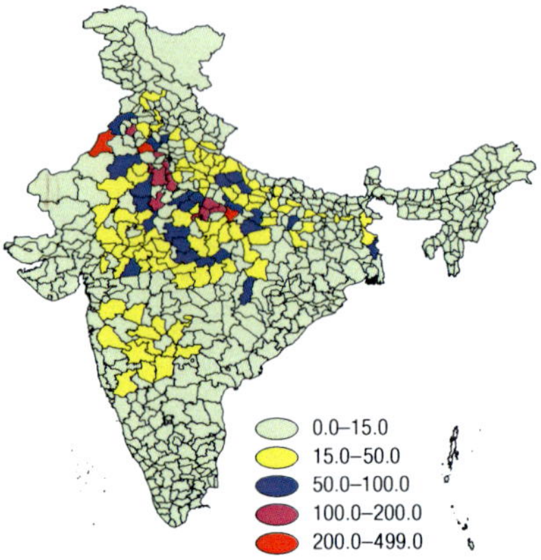

Chickpea production ('000 tons) in 1970

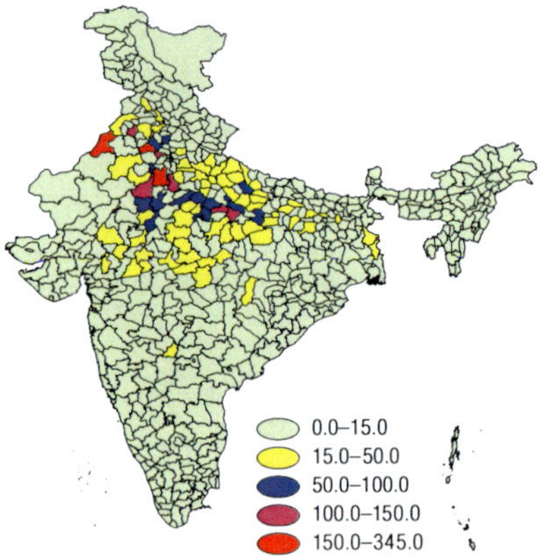

Chickpea area ('000 Ha) in 2007

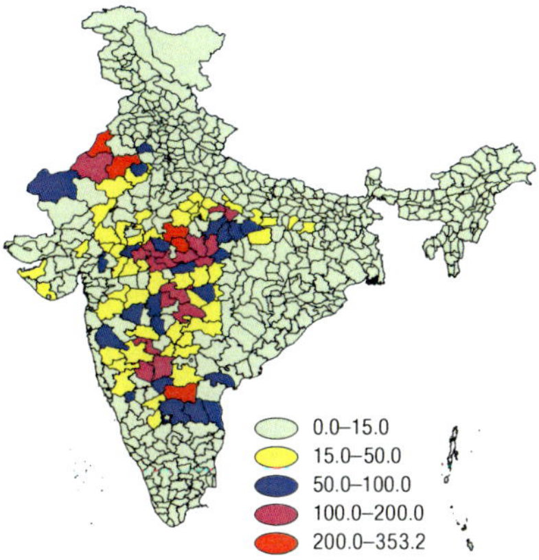

Chickpea production ('000 tons) in 2007

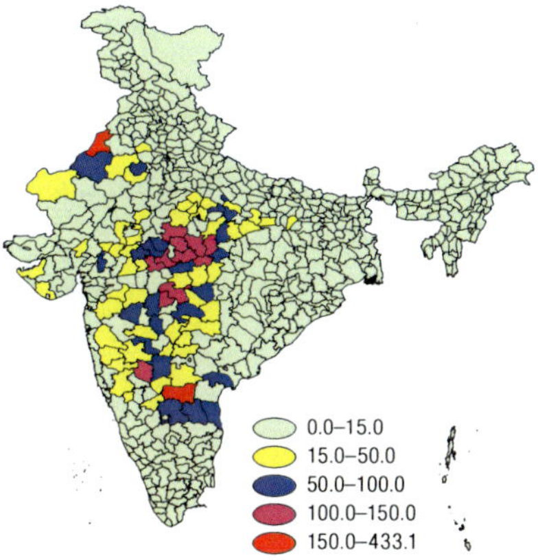

FIGURE 3A.4 Shift in pigeon pea area and production across states, 1970 and 2007

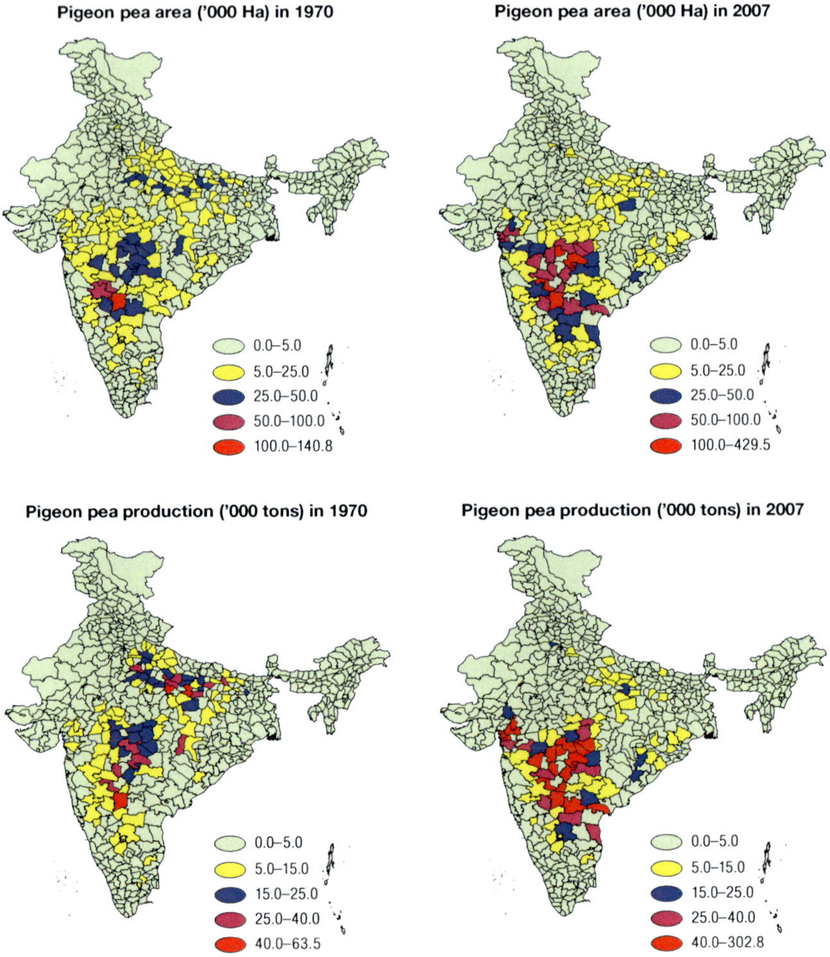

Source: Authors' calculations based on data from India, Directorate of Economics and Statistics, Ministry of Agriculture, various years.

FIGURE 3A.5 Grouping of different states of India based on area allocation and production in chickpea

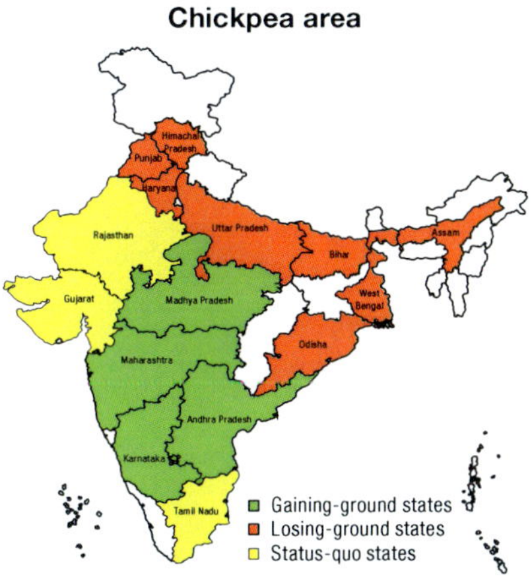

Source: Authors' calculations based on data from India, Directorate of Economics and Statistics, Ministry of Agriculture.

FIGURE 3A.6 Grouping of different states of India based on area allocation and production of pigeon pea

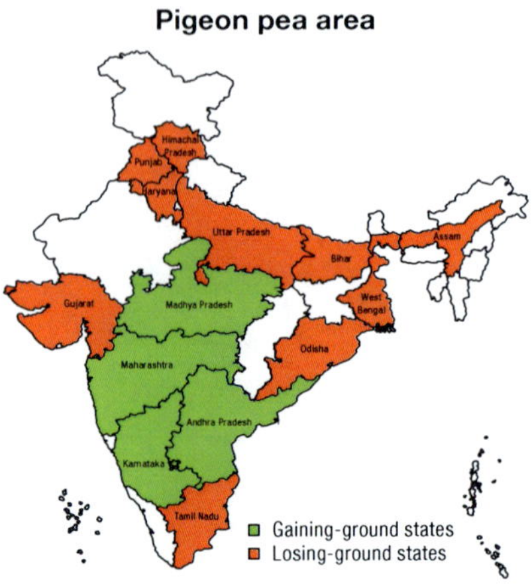

Source: Authors' calculations based on data based on India, Directorate of Economics and Statistics, Ministry of Agriculture.

Chapter 4

TECHNOLOGICAL INNOVATIONS IN PULSE PRODUCTION

B. Mishra

Not only do pulses contain higher protein in comparison to cereals, they also contribute to the sustainability of the environment through biological nitrogen fixation. For example, depending on the soil and agroecological environment, the chickpea plant fixes 23 kilograms to 97 kilograms (of nitrogen per hectare), pigeon pea fixes 4 kilograms to 200 kilograms, black gram fixes 119 kilograms to 140 kilograms, and green gram fixes 50 kilograms to 66 kilograms (Wani, Rupela, and Lee 1995). The biological nitrogen fixed by the preceding pulse crop saves a significant amount of synthetic nitrogen input that would otherwise be needed by the subsequent crop, carrying potential risks to the environment due to nitrogen runoff contaminating underground and surface water. This combination of high nutritional value and environmental benefit, together with their established place in traditional Indian diets, makes pulses an important target of long-standing and continuing agronomic research.

Background

Research on pulses in India started as early as 1905, with a modest beginning at the Imperial Agricultural Research Institute, now known as the Indian Agricultural Research Institute (IARI). Among pulses, chickpea and pigeon pea have a long history of research in the states of Madhya Pradesh, Maharashtra, and Uttar Pradesh. Research work on green gram and black matpe was initiated at Pusa in Bihar in 1925 with the collection of landraces (the varieties traditionally cultivated over a very long period) and selection based on various traits. However, systematic research on pulses commenced only with the establishment in 1967 of the All India Coordinated Pulses Improvement Project (AICPIP), which was later elevated to the Indian Institute of Pulses Research (IIPR). Table 4.1 presents the facts about All India Coordinated Research Project (AICRP) centers for different pulses in India. Assessed in terms of the number of centers, pigeon pea and chickpea

TABLE 4.1 All India Coordinated Research Project (AICRP) centers in India

Crop	Main center	Subcenter	Voluntary center	Total
Chickpea	9	15	34	58
Pigeon pea	9	17	11	37
Other	6	21	10	27

Source: AICRIP Annual Reports, various years

occupy the prominent position, with 75 percent of the centers focused on just these two pulse crops. The other pulses combined are covered by only 25 percent of the research centers. This chapter provides stylized facts about various technological innovations developed and adopted for pulse production in India over time.

Study Objectives and Data Sources

The objectives of the study presented this chapter are threefold: (1) to review the development of various technologies and innovations, including biotech innovations, in the production of the major pulses in India; (2) to look into the use of wild species as a source of gene pools, particularly for resistance to biotic and abiotic stresses; (3) to prepare an inventory of region-specific and environment-specific varieties of major pulses in India. The study is based on the secondary data collected from various publications, including the annual reports of the IIPR and the International Crop Research Institutes for the Semi-Arid Tropics (ICRISAT); the AICRP's reports on pigeon pea, chickpea, and other commodities; and several published research papers on the topic of technology development in pulses in India.

Development of Pulse Varieties

In India, pulse research has largely focused on five broad areas: (1) breaking the yield barrier; (2) developing resistance to pests and diseases; (3) breeding varieties for nontraditional and marginal areas; (4) reducing the length of the growing season by developing short-duration varieties; and (5) improving quality, especially in grain size. The early research efforts to break the yield barrier in pulses had limited success. Despite a growing demand-supply gap in pulses and modest progress on breaking the yield barrier compared to the achievements in rice and wheat, the number of full-time scientists engaged in

TABLE 4.2 Pulse cropwise varieties developed and released in India

	Number of varieties	
Pulse crop	Developed	Released (after 1990)
Chickpea	180	60
Pigeon pea	129	68
Green gram	130	45
Black matpe	90	34
Field pea	43	30
Lentil	45	26

Source: AICRIP Annual Reports, various years.

pulse research remains very low: at present, the ratio is only 2.5 full-time scientists per million hectares of pulse area (ASTI 2014–2015). In addition, there are scientists in the CGIAR centers, ICRISAT, and International Center for Agricultural Research in the Dry Area (ICARDA) who focus on pulses technology under agroclimatic conditions that are relevant to India.

Pulse research was unable to develop varieties to compete with the dwarf and high-yielding rice and wheat varieties that led to the Green Revolution. As described in detail in Chapter 3, since the 1970s pulses have in effect been largely thrown out of the Indo-Gangetic plains, supplanted by these high-yield cereals. Gradually but surely, pulses found new niches in the rainfed areas of the southern, western, and central parts of India. These nontraditional areas faced problems of drought, heat, pests, and diseases, so the challenge has been to fit pulses into new production systems under different agroecologies. Therefore, the varietal development programs in pulses have focused on selecting varieties for adaptation to the stresses of marginal environments rather than selecting for high-yield potential under unlimited conditions. The total number of varieties developed and released in the country for major pulses is quite large, as is evident in Table 4.2. Of course, development and release do not equate to adoption by farmers, an issue that is discussed in the next section.

Note that the social benefits of pulse R&D outweigh the private gains, because, except in the case of hybrids, private companies cannot fully appropriate the benefits of research into variety improvement, and the health and nutrition benefits of pulses are not likely to be fully reflected in market prices. Consequently, the private sector will likely underinvest in pulse varietal improvement research, and this market failure will persist. Persistent market failures necessitate adequate public investment.

Chickpea (*Cicer arietinum* L.)

In India, more than 180 chickpea varieties have been developed by the national program either alone or in partnership with international institutions. Since the 1970s, the focus of development has varied through different phases of the program. During the 1970s most of the varieties were developed through selection from landraces, with a major emphasis on increasing yield potential. During the 1980s the emphasis was on breeding to develop disease resistance. During the 1990s the major thrust was to develop varieties for multiple-disease resistance, stress tolerance, and high input response.

Through all of this work, the most significant breakthrough in India has been the development of short-duration chickpea varieties. A large array of short-duration varieties have been developed to handle different types of stresses and situations. Because these improved varieties are tolerant to heat stress, they have found a niche in central and peninsular India, where their adoption has worked well in the hot and dry climates. During the 1990s, in addition to developing disease resistance, stress tolerance, and high input response, genetic sources were deployed to breed varieties tolerant to drought, cold weather, and salt. As a result of the decade's work, a number of varieties were released that are resistant to wilt, to root rot, ... and to Ascochyta blight (Singh and Sewak 2013) (Table 4.3).

Regional adaptation. Several varieties resistant or tolerant to *Fusarium* wilt were developed and released for cultivation in different regions and states (Table 4.4). These varieties were also high-yielding compared to the local landraces, and some were well suited to growing in nontraditional areas as well. Similarly, varieties tolerant to *Ascochyta* blight were developed for the country's Northwest Plain zone (especially Punjab, Haryana, northwestern Rajasthan, and western Uttar Pradesh). According to the Agriculture Science and Technology Indicators (ASTI) (https://www.asti.cgiar.org/trivsa), the rate of adoption of improved chickpea varieties in select Indian states in 2010 is as follows: Andhra Pradesh, 99 percent; Karnataka, 100 percent; Madhya Pradesh, 84 percent; Rajasthan, 68 percent; and Uttar Pradesh, 65 percent.

As these improved varieties expanded into new areas, they also replaced the traditional varieties. By 1995 about 52 percent of the existing chickpea area was allocated to improved varieties in Andhra Pradesh, Gujarat, and Madhya Pradesh (Joshi, Asokan, and Bantilan 1999). The result was substantial gains in both chickpea yields and farmers' incomes. The yield advantage was

TABLE 4.3 Environment-specific chickpea varieties at a glance

Environment	Varieties
Short-duration	ICCV2, JG74, Vijay, JG11, JG16, JAK1, 9218, KAK2
Salt-tolerance	CSG 8962 (Karnal Chana 1)
Drought-tolerance	ICCV10, Phule G5, RSG 888, Vijay
High-yielding kabuli chana varieties	KAK2, BG1003, BG1053, Phule G 95311, IPCK 2002–29

Source: AICRIP Annual Reports, various years.

reported to range from 28 percent in Andhra Pradesh to 67 percent in Gujarat. Moreover, the yield gains were much higher for high-yielding bold (*kabuli*) varieties, ranging from 108 percent in Andhra Pradesh to 123 percent in Madhya Pradesh (Joshi, Asokan, and Bantilan 1999). The farmers also benefited from the price premium they gained in the market due to the new varieties' size, color, and shape (Shiyani et al. 2002). The study reported that a silent "chickpea revolution" was witnessed in central and peninsular India, although it was combined with a gradual decline in chickpea area and production from northern and eastern India. Most notably, the production of chickpea in hot and dry climatic regions, which had contributed 40 percent of the country's total in the early 1980s, increased to account for 75 percent of total production in 1995 and 86 percent as of 2013–2014.

Adoption and yield. A recent study of the adoption of improved chickpea varieties in southern India's major chickpea-growing areas found that 97 percent of the farmers were adopting improved varieties, covering nearly 86 percent of the chickpea-growing area (Suhasini et al. 2012). A comprehensive 2014 study found that in Andhra Pradesh, as of 2011, nearly 90 percent of the area under chickpea production was planted in improved varieties (Bantilan et al. 2014). These findings testify to a high rate of seed replacement, with government figures showing that improved varieties expanded from a mere 3 percent in share in 2001 to 85 percent in 2011 (India, Ministry of Agriculture 2016). The comprehensive 2014 study estimated the direct welfare gains from investing in chickpea research and adopting the improved varieties to be on the order of US$358.9 million in Andhra Pradesh alone, before accounting for the general equilibrium effects, and particularly the price reduction, that would result from wider adoption of the improved variety across India.[1]

1 All dollar figures used in the chapter are US dollars.

A large share of the short-term gains (99 percent) accrued to the adopting farmers in Andhra Pradesh (Bantilan et al. 2014). Once general equilibrium effects were taken into account, the same study estimated the total net contribution of chickpea research investment at the all-India level to be \$543.9 million (at constant prices), of which 85 percent (\$450.2 million) would accrue to consumers (benefitting from price reduction), and 15 percent (\$93.7 million) would accrue to adopting farmers across India. Nonadopting farmers would sustain a significant welfare loss. The overall internal rate of return on the investment in chickpea research was estimated at 28 percent.

New technologies. In addition to varietal improvement, new technologies introduced for chickpea production have included integrated pest management (IPM) and the development of farm machinery. IPM is discussed in greater detail later in this chapter. Concerning mechanization, the first machine-harvestable chickpea variety (NBeG 47) was recently released in Andhra Pradesh to overcome the problems of labor shortage and high wages (ICRISAT 2016). The machine can harvest 2.25 tons in 75 minutes, a boon for chickpea production in labor-scarce areas.[2]

Pigeon Pea (*Cajanus cajan* [L.] Millspaugh)

Pigeon pea is a long-duration and indeterminate crop that is prone to numerous diseases and insects and also suffers from low yield. The focus of research therefore has been fourfold: (1) to develop varieties of medium to short duration without compromising yield levels; (2) to develop resistance to pests and diseases; (3) to develop determinate varieties for uniform crop maturity; and (4) to increase yield levels. A number of varieties with different traits to handle a wide range of environments were developed, and a few promising ones are listed in Table 4.4.

Duration. Traditionally, the pigeon pea crop matured in 280 to 300 days, and in some cases ratooning was done for two to three years.[3] Therefore, developing short- and medium-duration varieties became a high research priority. During the 1980s and early 1990s, several medium- and short-duration varieties were developed, some of which found new niches and were adopted

2 All measurements in tons in this chapter are in metric tons.

3 Ratoon cropping is a multiple-harvest system in which regenerating stubbles of the established crop in the field are managed for subsequent production. The development of short-duration varieties of pulses has generated interest in ratooning. Instead of cutting whole plants, only the pods are picked from the ratoon crop and the plants are allowed to bear their next flush of pods. Irrigation after the main harvesting of the crop increases the yield from the ratoon crop (see Bantilan et al. 2014).

TABLE 4.4 Environment-specific varieties of pigeon pea at a glance

Environment type	Varieties
Wilt resistance	Maruthi, Asha, BDN2, BSMR 736, MA 6
SMD resistance	Bahar, BSMR 736, Asha, Sharad, Pusa 9
Wilt and SMD resistance	Asha, BSMR 736, BSMR 853
Hybrids	ICPH8, PPH4, COPH1, COPH2, AKPH410, AKPH2022

Source: AICRIP Annual Reports, various years.

in nontraditional areas. These varieties matured in 140 to 160 days without compromising the yield levels. Some of these varieties led to diversification in the rice-wheat production systems in northern India. Important among these varieties is the pea known as UPAS 120, which is the most popular. Its yield is in the range of 1.6 tons to 2.0 tons per hectare and it matures in just 120 days. It is most suitable for double cropping. Some estimates suggest that such short-duration varieties have expanded the pigeon pea area in northern and northwest India by roughly 200,000 hectares. The falling water table and remunerative pigeon pea prices are believed to be the leading factors motivating farmers to adopt these varieties.

A downside to the first available medium- and short-duration varieties is that they were susceptible to a few diseases (such as sterility mosaic, *fusarium* wilt, and *phytophthora* blight) and tended to prolong their maturity into the late monsoon rains. A breeding program, therefore, focused on developing varieties that would mature by early November to escape these diseases and fit well into the multicrop production system to ensure the timely sowing of wheat. Among other attributes, such as their determinate growth habit, short stature, and early maturity (120–130 days), several of the resulting new cultivars (like ICPL 87) proved to be suitable for both sole cropping and multiple harvesting.

The ICPL 87 variety was an especially successful example. It emerged from the National Pulse Development Program for Western Maharashtra, where sustainability of water and soil were adversely affected by the cultivation of sugarcane and banana. Designed to thrive in an irrigated environment in rotation with other crops, the variety also offered several other advantages, including enhancing income, improving soil health, and adapting to drought stress. By the mid-1990s it had been adopted across all districts with access to irrigation in Western Maharashtra (Bantilan and Parthasarathy 1999).

Disease resistance. Wilt is one of the major diseases that seriously harmed pigeon pea yield in earlier decades. Globally, estimates showed that wilt reduced yields by up to 50 percent (Ryan 1981). Wilt-related production losses in 1977–1978 were estimated to be about $36.4 million in India and $5.2 million in Kenya, Malawi, and Tanzania (Ryan 1981). Research efforts yielded several wilt-resistant varieties for India and Africa. Among others, ICP 8863 was widely adopted in the semiarid tropics. Adoption studies in India revealed that this variety occupied almost 60 percent of the pigeon pea area in the wilt-affected districts of northern Karnataka and the bordering districts of Andhra Pradesh and Maharashtra. The potential benefits of adopting the wilt-resistance variety were estimated to be $79.8 million (Bantilan and Joshi 1996).

Adoption and yield. Adoption and impact assessment of pigeon pea varieties has not received due attention from professionals. However, one evaluation by ICRISAT has assessed the impact of improved pigeon pea varieties in rainfed areas of Odisha (Mula et al. 2014). The study found that the improved varieties had higher grain yields (70 percent) when compared with landraces. It also found significant increases in net income (which rose from 170 percent to 190 percent) and greater participation of women farmers (34 percent) in production. Unfortunately, the study did not trace the adoption of improved pigeon pea varieties in the study locations. According to an ASTI study in 2010 (https://www.asti.cgiar.org/trivsa), the rates of adoption of improved pigeon pea varieties in some Indian states is as follows: Andhra Pradesh, 70 percent; Maharashtra, 70 percent; Tamil Nadu, 70 percent; Madhya Pradesh, 65 percent; and Uttar Pradesh, 25 percent.

An additional program to break the yield barrier was an effort in India to develop a hybrid pigeon pea, which resulted in the world's first pigeon pea hybrid (ICPH 8), released in 1991. This hybrid was of short duration, offered a high-yield potential, and was drought-tolerant. Since then, the successful development of hybrids has opened up new avenues for enhancing the yield potential in pigeon pea (Saxena et al. 2005; Saxena 2009). Extensive testing of pigeon pea hybrids has shown yield advantages of 40 percent to 47 percent over the local varieties and even over other improved varieties in farmers' fields in India (Saxena and Nadarajan 2010).

In trials of hybrids conducted in five states, the mean yield (1,396 kilograms per hectare) was 47 percent higher than the yield of a popular variety (ICP 8863, 953 kilograms per hectare). The hybrids also exhibited high levels of resistance to the *Fusarium* wilt and sterility mosaic diseases (Saxena et al. 2013). Two of the recently released hybrids (ICPH 2740 and ICPH 14003) possess resistance to wilt and sterility mosaic diseases in Andhra Pradesh. These hybrids have a high-yield potential of 2.5 tons to 3.5 tons per hectare, which is 25 percent to 40 percent higher than the local varieties. However, although several of these hybrids have been released for cultivation, they have not been particularly successful at getting adopted. Four major constraints to their adoption have been documented: (1) the high labor cost for seed production; (2) the high seed rate (amount of seed sown per hectare); (3) heavy damage from pod borers; and (4) lack of knowledge among farmers about seed production (Niranjan et al. 1998).

New developments. The next generation of breeding and agronomic efforts in pigeon pea will be focused on improving the plant type. Unfortunately, the genetic base of pigeon pea is quite narrow, with only 57 ancestors having been used for the development of 47 varieties through hybridization following selection. Only 32 wild species are known as valuable sources for resistance or tolerance to several biotic and abiotic stresses. But only 1 percent of the entire collection has actually been used to identify the sources of resistance to diseases, drought, and other abiotic stresses (Upadhyaya et al. 2009). Scientists are now using specific attributes, such as determinate growth habit, short stature, and early maturity (120–130 days), to develop varieties suitable for sole cropping and single or multiple harvesting. Research on developing transgenic varieties is now at an advanced stage, and pigeon pea is amenable to genetic transformation using recombinant DNA and tissue culture. Effective protocols are available to carry this regeneration out through organogenesis and somatic organogenesis in pigeon pea. Transgenic plants with the *Bt* gene have been tried for imparting resistance against *lepidopteron* insect pests that affect pigeon pea, but there is no product as yet, and we think that research on this needs to be systematized and intensified.

Green Gram (*Vigna radiata* L. Wilczek)

Green gram is mainly a rainy-season crop, although it is also grown during winter and summer. Its varietal development program has largely occurred in four phases with four areas of focus: (1) increasing yield potential; (2) reducing the duration; (3) developing resistance against diseases (especially powdery mildew and mosaic virus); and (4) breeding for large grain size.

Disease resistance. During the 1970s, research efforts mainly employed hybridization and mutation to breed high-yielding varieties. In the 1980s hybridization was used widely to combine agronomically useful traits and disease resistance. This led to the development and release of several varieties resistant or tolerant to powdery mildew and mung bean yellow mosaic virus (MYMV). During the 1990s, several sources of large grain size (> 6 grams per

100-seed weight) were introduced from the World Vegetable Center (formerly known as Asian Vegetable Research and Development Center, or AVRDC) and widely used in the Indian breeding program. According to the Asian Vegetable Research and Development Center (AVRDC 1998), MYMV is a virus with the potential to cause crop losses as high as 85 percent.[4]

By the mid-1990s, several large grain size varieties were developed and released, as were several multitrait varieties with desirable properties, including large grain size, short duration, photo-thermo insensitivity (resistant to heat and excessive sun), synchronous maturity, and resistance to major diseases. Recently, the incidence of MYMV disease has become a serious problem in the rice fallows of south India, so efforts have been diverted toward incorporating MYMV-resistant genes along with powdery mildew resistance. Some of the ruling varieties are listed in Table 4.5, along with their salient traits.

Broadening the genetic base. It is unfortunate that only a limited amount of the genetic variability in green gram has been exploited in varietal development programs. There is enormous potential to use known wild species and cultivate *Vigna* species to incorporate novel traits and broaden the genetic base. The gene introgression in green gram has already resulted in green gram derivatives that have shown potential for raising yields and building disease resistance. Moreover, these derivatives facilitate further genetic enhancement in green gram. The efforts made in this direction have led to the development of several improved cultivars of green gram, such as IPM 99-125, IPM 02-3, and IPM 02-14 (Singh, Dixit, and Katiyar 2010).

Current research. To further boost yield and find new niches, researchers are looking for ways to substantially change the plant's architecture. The available plant type in green gram is largely photo-thermo sensitive, with an indeterminate growth habit, low harvest index, and low grain yield. To remain a commercially competitive crop, green gram will have to fit into the production cycle of an intensive-input cereal-based cropping system. Therefore, the direction of future research is to develop a plant type that is determinate, photo-thermo insensitive, early maturing, high yielding (1.5–2.0 tons per hectare), with a high harvest index, and resistant to lodging and diseases. There is also a need to develop varieties of varying duration for India's different agroclimatic zones.

---

4 The AVRDC is now called the World Vegetable Center.

TABLE 4.5 Environment-specific varieties of green gram at a glance

Attributes	Varieties
Short-duration varieties for spring/summer	IPM 02–3, Meha, Samrat, TMB 37, HUM 16, HUM 1, PusaVishal, OUM 11–5, Pant M 5, SML 668
Powdery mildew–resistant varieties for *rabi* season	TARM 18, TM 96–2, Vamban 2, Vamban 4, TARM 2, TARM 1
Mung bean yellow mosaic virus–resistant varieties	Pant M 4, Pant M 6, KM 2241, Sattya, NDM 1, HUM 1, Ganga 8, Samrat, Meha, HUM 12, IPM 02–3
Large-seeded (5 grams/100 seeds)	Pant M 5, Pusa Vishal, SML 668, HUM 16, TMB 37, IPM 02–3

Source: AICRP Annual Reports (various years).

Black Matpe (*Vigna mungo* L. Hepper)

Black matpe is widely grown on the Indian subcontinent, where it originated and has been cultivated since ancient times. Early research on black matpe was started in the 1940s. The initial research phase focused on varietal development for improving locally adapted but genetically variable populations, mainly pure line and mass selections with a major emphasis on traits rather than yield. This resulted in the release of a large number of pure lines, some of which are still cultivated in certain parts of the country. Between 1943 and 1953 a large number of black matpe varieties (including T 27, T 77, and T 9) were developed that were adapted to the northern Indian environment. The variety T 9 is still a very popular variety, which is a striking fact considering how long ago it was developed.

Before 1970, several varieties of black matpe were developed from the locally adapted varieties. The most important lines preferred by farmers and most extensively used in the breeding programs were T 9, ADT 1, and CO 1. In the late 1970s, a hybridization program was started to develop short-duration and MYMV-resistant varieties. KM 1, which appeared in 1977, was the first variety developed through this hybridization. Later, a large number of varieties were developed for different ecosystems with varying traits using hybridization.

Disease resistance. In the 1980s the research priority for black matpe was to develop disease-resistant varieties. At that time, powdery mildew was the major disease of concern. The first variety resistant to powdery mildew was LBG 17, which was developed and released in 1983 for the rice fallow systems of coastal areas. It was so successful in its adoption that it revolutionized black matpe cultivation in the coastal regions of Andhra Pradesh. Later, more varieties of black matpe were developed and released, which led to an expansion in their growing area in rice fallow land of the coastal peninsula. Since 1990, the major emphasis has been on breeding short-duration, photo- and thermo-insensitive varieties of black matpe along with resistance to biotic stresses (namely, yellow mosaic virus and powdery mildew). Some of the important black matpe varieties are given in Table 4.6.

Current research. As with other crops discussed previously, the varietal development program for black matpe has only exploited a limited amount of the plant's potential variability. The variety T 9 alone has contributed 75 percent to the development of new black matpe varieties. The genetic base of the available varieties is very narrow, but there is scope for utilizing the available gene pools from wild species to broaden the genetic base and borrow novel traits. Gene introgression in black matpe has already resulted in derivatives that have shown potential for yield-contributing traits and disease resistance. These derivatives have facilitated further genetic enhancement in black matpe, which has led to the development of improved cultivars like Mash 1008 and VBN 5 in black matpe (Singh, Dixit, and Katiyar 2010). These have facilitated the cultivation of black matpe in diverse agroecological regions.

Lentil (*Lens culimaris* Medic)

Lentil is gaining popularity due to its rich nutritional value; it is high in protein and iron and has no fat or cholesterol. In India today about 1.5 million hectares are planted in lentil. About 85 percent of lentil is produced in only three states—namely, Bihar, Madhya Pradesh, and Uttar Pradesh. The research efforts in lentil have been lackluster, particularly until the 1990s, and confined mainly to identifying landraces for better adaptability and yields. During the 1980s a few landraces were collected and used in recombination breeding through single crosses, followed in the 1990s by crosses involving more parents. The narrow genetic base has also been used to breed high-yielding, short-duration, and disease-resistant varieties, including many good varieties resistant against rust, *fusarium* wilt, and vascular wilt. Four high-yielding varieties resulted that have been commercialized—namely, Angoori, Noori, Priya, and Sheri. In the early 1990s, International Center for Agricultural Research in Dry Areas (ICARDA) introduced an early flowering line (Precoz: ILL 4605) that was used in hybridization with indigenous lines and has resulted in the selection of extra-early genotypes. ICARDA is providing India's national program with valuable nurseries, which are used for fixed and segregating populations for various desirable traits. Several varieties have been developed and released for cultivation using the source material received from ICARDA (Sarkar et al. 2007).

TABLE 4.6 Environment-specific varieties of black matpe at a glance

Attributes	Varieties
Short-duration varieties for spring/summer	WBU 109, Azad Urd 1, KU 300, Pant Urd 31, PDU 1, KU 92–1
Powdery mildew–resistance	LBG 625, LBG 685, LBG 623, LBG 20, WBG 26, LBG 709, LBG 645, VBN 4
Mung bean yellow mosaic virus–resistance	WBU 108, Pant U 30, Pant U 31, Pant U 40, Azad U 1, Azad U 2, Sekhar 2, Sekhar 3, IPU 02–43, Uttara, NDU 1, KU 96–3, Mash 1008, WBU 109
(Mung bean yellow mosaic virus + powdery mildew)-resistance	IPU 02–43, LBG 625, LBG 685

Source: AICRP Annual Reports (various years).

The improved varieties in lentil have made it possible to achieve wide adaptability in varying agroenvironments. The short-duration varieties that have been made available fit well into any production system where there is residual moisture. Over a period of four decades, the varietal development has been reflected in increased lentil production, which rose from 0.37 million tons in 1970–1971 to 1.13 million tons in 2012–2013. Both yield increase and area expansion contributed to this jump. National average yield increased over the same period from a mere 497 kilograms per hectare to around 800 kilograms per hectare, while in Bihar yield rose to more than 1,100 kilograms per hectare. Lentil area also doubled over the past four decades due to the availability of varieties that are disease-resistant and short-duration, expanding from 0.75 million hectares to 1.42 million hectares.

Transferring the Technology to Pulse Farmers

Pulses are seriously affected by a large number of insects and diseases. Important among these are the pod borer (*Helicovera armigera*) followed by the pod fly, wilt, and root rot. It is reported that *Bacillus thuringiensis* Berliner (Bt) var. *kurstaki* is effective in controlling pod borer; however, the successful release of Bt chickpea/pigeon pea varieties from either public or private research will take several more years. The other important pests affecting pulses are nematodes, among which root-knot nematodes are important due to the way they spread and damage crop yield. Among important diseases, wilt in chickpea, sterility mosaic virus (SMD) in pigeon pea, and yellow mosaic virus (MYMV) and powdery mildew (PM) in green gram and black matpe cause major damages to pulse crops.

Various combinations of the integrated pest management (IPM) approach have been developed, piloted, and disseminated for controlling pests and diseases in pulses. The principle of IPM is to minimize the application of chemicals and manage pests and diseases through better crop management. Considerable efforts were made to promote IPM, although they have not succeeded as expected. The main constraint has been that IPM practices require collective action among farmers.

Production of Improved Varieties of Pulses

The availability of improved-quality seeds is one of the most important drivers for increasing pulse production, especially for resource-constrained farmers in rainfed areas. Unfortunately, the scarcity of breeder's seed and certified seed of improved varieties is constraining their adoption. There is a weak link between the research and development systems and the mass production and multiplication of breeder, foundation, and certified seeds. The link of the seed system with the market in terms of the desired varieties is also weak. The existing seed sector is fragile because of the low volume of business. However, the seed replacement rate of important pulses is increasing somewhat in recent times, although it is still quite low. For example, in chickpea the seed replacement rate is merely 14 percent (India, Ministry of Agriculture 2016). Gowda et al. (2013) point to several factors for low seed replacement rates such as low seed multiplication rate of legumes; reuse of grains from the previous harvest as seeds; and often demand for specific varieties adapted to more narrow agroecologies and consumers' needs.

Despite a long list of improved pulses varieties released for cultivation, their impact has not yet been fully realized by the resource-poor farmers in many states in India. The accessibility of smallholders to quality seed of improved pulses varieties is constrained by both inadequate demand creation and limited supply (Gowda et al. 2013). The policy support for the pulses seed system has been unfavorable and regulatory frameworks have been inadequate (Rubyogo, Sperling, and Assefa 2007).

Gowda et al. (2013) point out that legume seed business in general in India does not attract large seed companies since profit margins are low. In this context, Materne and Reddy (2007) point out that more than 95 percent of lentil seed in India (the world's largest producer of lentil) is sourced from the informal sector and this type of dominance of informal sector is quite generic for the pulses sector as a whole. The formal seed sector is still concentrated in areas with high population density and areas with better infrastructure. The small and medium seed companies that are emerging still have limited capacities and apart from marketing problems lack a good supply of foundation seeds (Gowda et al. 2013).

Going forward, for the pulses seed sector, Singh and Saxena (2016) suggest several options to ensure quality pulses seed availability in the country. These are

- Adoption and promotion of new varieties by bringing them into the seed chain by the state departments.
- Production of Truthfully Level seed (TL seed) at research institutions and state agricultural universities (SAUs) as well as at Krishi Vigyan Kendra (KVK) or Agriculture Science Centers farms and their distribution to farmers.
- One cycle multiplication of certified seed at KVK farms before their demonstration/distribution to farmers.
- Continuation of subsidies for additional years for popular varieties that have large seed indent (requirement) but are going to be phased out of the seed chain due to completion of time period.[5]
- Strengthening infrastructures of research farms and KVKs for increasing seed multiplication ratios and developing/strengthening seed processing and storage facilities.

Singh and Saxena (2016) further argue that in the case of breeder seed production, some of the varieties are quite old and need to be gradually substituted by new varieties. Among the chickpea varieties—JG 11, JG 16, JG 322, and Vijay—there are more than 15 years of continuous subsidies among them. Similar is the case of pigeon pea and green gram. Singh and Saxena (2016) document that the average seed replacement rate (SRR) of pulses in India was about 25 percent at the end of 2011. The highest SRR was in the case of black matpe (34.41 percent), followed by green gram (30.29 percent), and pigeon pea (22.16 percent). Singh and Saxena (2016) like Gowda et al. (2013) point out that among major production constraints, availability of quality seed of

5 Indent is a seed requirement by different states given to the central government. Each state assesses its own seed requirement and gives indent to the central government. For just-released varieties, indent of breeder seed is given to Indian Council of Agricultural Research/SAUs and for old varieties indent for certified seed is given. There is a committee at the national level, which meets twice a year for this purpose for all crops.

improved varieties has been a major constraint in enhancing production and productivity of pulses in India.

This they relate primarily to lack of an organized seed production program for pulses where there still is a lack of proper medium-term (four to five years) seed rolling plan for the country's major pulse-producing states. The indent for breeder seed is quite low in many cases, including indent for old and obsolete varieties. Moreover, there is poor conversion of breeder seed to foundation and certified seed. To ensure timely availability of quality seed, capabilities of seed production must be enhanced with multiagency participation, such as seed societies, farmers, private sectors, and NGOs besides SAUs, Indian Institute of Pulses Research, and State Seed Corporations (Singh and Saxena 2016).

It is necessary to improve the seed system in providing quality seeds of the improved and preferred varieties in adequate quantities and at affordable prices, at the right place and the right time. There is also a need to find ways to link the formal and informal seed sectors to achieve sustainable seed delivery to farmers and to explore approaches to motivate small and medium-size seed companies and NGOs to enter the pulse seed production program. The public seed corporation should increase seed production to meet the future target of national demand.

Technology-Farmer Link

Currently, there is a significant unexploited potential of pulses in terms of yields. This is demonstrated in Figure 4.1, derived from Singh and Saxena (2016), which shows existing yields of different types of pulses in India alongside the yields achieved at experimental stations and in field trials. The yield gaps are quite significant, ranging from 75 percent in lentil to 224 percent in green gram. Singh and Saxena (2016) suggest that the underlying reasons for these gaps are mainly poor quality of seed and poor management practices.

To strengthen technology transfer to farmers and acquaint them with new varieties and management practices, in 1990–1991 the Indian Council of Agricultural Research inaugurated its Front Line Demonstrations (FLD). The FLDs are helpful to both researchers and producers of pulses, but their target groups are farmers and extension service delivery workers. The demonstrations are conducted with farmers under the close supervision of scientists on a block of 2 to 4 hectares, where the latest and most promising pulse varieties and management practices are exhibited. Participating farmers are trained in the complete package of practices necessary to attain the potential yields. The approach also allows for farmers and extension workers to provide feedback,

FIGURE 4.1 Actual and potential yields in different pulses (kilograms per hectare)

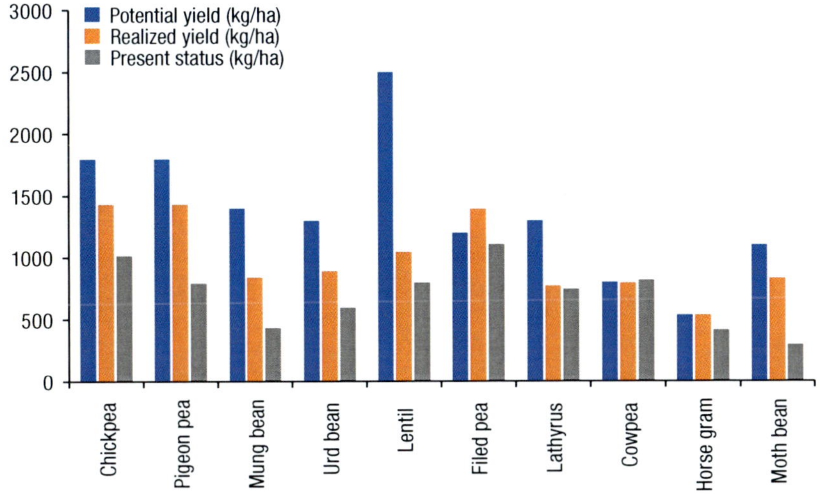

Source: Singh and Saxena (2016).

and it generates evidence that researchers can use to identify what factors are contributing to higher crop yields and what the constraints are under different farming situations.

The crops covered under the scheme are chickpea, pigeon pea, green gram, black matpe, field pea, lentil, kidney bean, lathyrus, and arid legumes (cowpea, guar, moth bean, and horse gram). The scheme over time has been transformed into a new mission mode program, Integrated Scheme on Oilseeds, Pulses, Oil Palm, and Maize (ISOPOM), with the following objectives:

- To demonstrate newly released crop varieties, their production and protection technologies, and their management practices at farmers' fields under different farming situations.
- To study the factors contributing to higher crop production.
- To generate production data and feedback information.

From 1996 to 2001 the field demonstrations showed the efficiency of improved technologies in enhancing the productivity of green gram, black matpe, lentil, field pea, kidney bean, and lathyrus. A total of 324 on-farm demonstrations were organized by various centers to test different technology components, including varieties. They found that pulse productivity could be increased by 50 percent to 100 percent (FLDs [1996–2001], AICRP 2013). From 2002 to 2007, 1,252 FLDs were conducted on chickpea in the major chickpea-growing states under the AICRP. The yield gains from improved varieties, measured in comparison with the yields of local varieties, ranged from 8.7 percent in Karnataka to 29.7 percent in Madhya Pradesh, with an average increase for all demonstrations of about 20.3 percent. From 2007 to 2013 the number of FLDs in chickpea and pigeon pea in different states was increased to 2,891 and the overall increase in yield due to improved varieties over local ones was 18.6 percent. The highest increase in yield was recorded in the state of Chhattisgarh (45.7 percent), followed by Uttar Pradesh (36.4 percent), Madhya Pradesh (25.5 percent), Maharashtra (22.3 percent), Gujarat (21.3 percent), Rajasthan (15.5 percent), Karnataka (10.9 percent), and Andhra Pradesh (9.9 percent).

From 2012 to 2013 the FLDs revealed that even higher average increases in yield (over time) were possible with farmers' adoption of a complete package of improved technologies, including an increase of 26.5 percent in chickpea and 30.4 percent in pigeon pea. Similarly, for other pulses, a higher grain yield was achieved through the adoption of a package of improved technologies, with increases of 22.9 percent in *kharif* green gram, 32.0 percent in *rabi* green gram, 29.5 percent in *kharif* black matpe, 29.4 percent in *rabi* black matpe, 21.0 percent in lentil, and 15.0 percent in field pea (AICRP 2013). One should note that the results from experimental plots of the kind promoted by the initiatives reported above are potentially biased upward in terms of yield improvement because they do not incorporate farmers' real-world constraints related to labor, credit, and so on. It is also not known what the impacts of the initiatives were, in terms of increased adoption in the neighboring areas.

Potential and Niche Areas for Pulses

Pulses are finding new niches as a result of the availability of improved varieties suited to new geographic areas. Table 4.7 shows that a minimum of 3.35 million hectares in India can potentially be used for cultivation of pulses (more recent estimates of rice fallow lands based on India, Ministry of Agriculture [2016] They are show the area of rice fallow lands equal to 4 million hectares). The discussion here highlights that the extent of technological progress in pulses has largely been a function of the ability to use fallow lands. According to NAAS (2013), rice fallows are found in the states of Andhra Pradesh, Assam, Bihar, Chhattisgarh, Jharkhand, Madhya Pradesh, Odisha, West Bengal, and Uttar

TABLE 4.7 Potential niches for pulses

Cropping system	Potential niches	Potential area (in millions of hectares)	Suitable varieties of pulse crops
Pigeon pea–wheat	Haryana, Punjab, northwest Uttar Pradesh, and north Rajasthan	1	UPAS 120, Manak, Pusa 33, AL 15, AL 201
Maize–*rabi* pigeon pea	Central and eastern Uttar Pradesh, north Bihar, West Bengal, Assam	0.3	Pusa 9, Sharad
Maize–potato/ mustard + green gram/black matpe	Punjab, Haryana, and west Uttar Pradesh	1	Green gram: Pant Mung 2, PDM 11, HUM 2, SML 668, Pusa Vishal; black matpe: PDU 1, Narendra Urd 1, Uttara
Spring Sugarcane + green gram/black matpe	East Uttar Pradesh, Bihar, West Bengal	0.15	Green gram: Pant Mung 2, PDM 11, Narendra mung 1; black matpe: PDU 1, Pant U 19, TARM 1, Pusa 9072
Rice–green gram	Odisha, parts of Karnataka, Tamil Nadu, Andhra Pradesh	0.35	TARM 1, Pusa 9072
Rice–black matpe	Coastal areas of Andhra Pradesh, Karnataka, Tamil Nadu	0.35	LBG 17, LBG 402
Rice–wheat–green gram	Western Uttar Pradesh, Haryana, Punjab	0.1	Pant Mung 2, Narendra, Mung 1, PDM 139, HUM 2
Maize–kidney bean–green gram	Central and eastern Uttar Pradesh, north Bihar	0.07	Green gram: Pant Mung 2, PDM 11, HUM 2; kidney bean: HUR 137, HUR 15, PDR 14, Amber
Kidney bean–Potato	Eastern and central Uttar Pradesh	0.03	PDR 14, Amber

Source: Ali (2004).

Pradesh. In these areas, after the harvest of *kharif* rice, climatic conditions of rice fallow lands are suitable for growing cool and warm season pulses by using the residual moisture. The pulses that can fit into these rice fallow systems are lentil, green gram, black matpe, lathyrus, and peas.

Pulses are finding new niches as a result of the availability of improved varieties suited to new geographic areas. Broadly, this process includes: (1) horizontal expansion into the rice fallow system in the coastal regions of Andhra Pradesh, Karnataka, Odisha, and Tamil Nadu; and (2) diversification within the rice-wheat system through the planting of short-duration green gram varieties and intercropping in sugarcane, pigeon pea, and cereals.

Naturally, such an expansion into a rice fallow system works best if the rice variety itself is of short duration and vacates the field early. Since lentil is more suitable and assured than chickpea in lowland areas with excessive soil moisture, lentil could be popularized in the lowlands of eastern Uttar Pradesh, Bihar, Jharkhand, and West Bengal (NAAS 2013). As explained in Chapter 3, Bihar, Jharkhand, and Odisha are also states where pulse cultivation has lost ground to other crops, so better utilization of rice fallow systems could help recover much of what was lost in these states. It is a fortunate coincidence that the scope for expansion in these states is comparatively large. Data shows that in the coastal areas of Odisha and Andhra Pradesh, as technologies were developed for rice fallow systems, an expansion in pulse cultivation occurred. Further use of rice fallow systems for pulse expansion in noncoastal areas, backed by the development of improved varieties of pulses and other technologies and by extension services, needs to be explored and prioritized. Indeed, a case may be made that expanding pulses into rice fallow systems should be tried first before attempting to extend the growth of pulses in new areas at the cost of competing crops like cereals or oilseeds.

The Way Forward

Pulse research needs to place a stronger emphasis on developing improved varieties that can break the existing low-yield barrier. Given the dynamics of supply in pulses in India, there is a need for horizontal expansion to new niches (such as the rice fallows in coastal regions of Andhra Pradesh, Karnataka, Odisha, and Tamil Nadu). Diversification in India's rice–wheat system has to be extended through short–duration green gram varieties and intercropping in sugarcane, pigeon pea, and cereals. Of high potential is the popularization of hybrid pigeon pea developed at International Crops Research Institute for the Semi-Arid Tropics (ICRISAT). Also, the progress made in using wild species for the introgression of valuable genes for their agronomic traits in different pulse crops—an approach that has largely been underutilized due to crossability barriers—can be scaled up. The following areas need greater attention in pulses research for technology development to meet the future demand.

Breaking the yield plateau and enhancing productivity. Three areas stand out as needing attention. First, the potential of biotechnology needs to be harnessed, including gene characterization for yield-determining traits using biparental populations, MAGIC (Multi-Parent Advanced Generation

Inter-Cross) populations, association mapping, and the development of functional markers for the genes. These research methods can be used without wading into the deeply divisive issue around commercializing genetically modified (GM) crops in India because these methods are not associated with the commercialization of GM crops. New tools of bioinformatics and statistical genetics should be used extensively, because this enables new genetic information to be generated very fast.[6] Second, the genetic base/gene pool needs to be widened. This includes prebreeding with wild pulse relatives. Third, hybrid technology needs to be developed with a suitable level of heterosis.

Developing crops resilient to climatic adversities. Resilient/smart pulse varieties and technologies need to be developed. Better monitoring of disease and pest dynamics in relation to climate change is needed as well.

Developing quality pulses. Due consideration must be paid to the quality traits when pulse varieties are identified for release.

Producing quality seed. Quality seed needs to be produced in sufficient quantity, effectively using the chain from breeder seeds to foundation and certified seeds.

Resource management. Ways to increase the input-use efficiency of nutrients and water need to be developed, with consideration for the differing macro and micro nutrients needed by different pulses and across environments. The efficiency of symbiotic processes for enhanced nitrogen fixation by Rhizobia needs to be improved. Finally, research is needed on microorganisms, like phosphate-solubilizing bacteria/fungi and biofertilizers such as vesicular arbuscular mycorrhiza (VAM), which are capable of solubilizing nonavailable phosphate into an available form and helping in phosphate uptake by pulse crops.

Conclusion

India has a long history of programs relating to technology development in pulses. Since pulses' preferences vary across regions, research has had to cover several varieties. The developments in pulse farming have faced several

6 The main objective of developing MAGIC populations is to promote intercrossing and shuffling of the genome. The advantages of using multiparent populations are that (1) more targeted traits from each of the parents can be analyzed based on the selection of parents used to make the multiparent crosses; and (2) increased precision and resolution can be detected due to the increased level of recombination (Cavanagh et al. 2008). Multiparent populations are now attractive for researchers due to the development of high-throughput genotyping platforms and advances in statistical methods to analyze data from these populations (see Bandillo et al. 2013).

dynamics that have had a significant bearing on technology development. First, pulses have moved significantly across regions. Technology has had to keep pace with these movements while adjusting to help the expansion across different regions (with different agroclimatic conditions) be successful. Efforts should be made to identify and map genes of economic importance from a large array of wild species and develop a molecular linkage map. Molecular markers linked to traits of agronomic importance in pigeon pea are limited. Microsatellite markers are being developed and mapped to overcome the relatively low amount of information that can be derived from the widely used dominant markers in pigeon pea.

Second, when the Green Revolution happened in cereals, it affected the prospects for pulses. Technology development had to focus not only on improving pulse yields but also on fitting pulse farming into the cropping complex that includes cereals. Hence, many short-duration and very-short-duration varieties had to be developed to meet the need to fit into rice-wheat cropping cycles. The increased focus on cereals meant that pulses have been increasingly pushed to marginal environments, creating a set of challenges to which technology development has had to adjust. Much of the technology development has focused on disease and pest resistance. Recent research that indicates the possible effectiveness of Bt-based technology in controlling pulses pests suggests directions for future research. The recent development of biopesticides could also be valuable in reducing the harmful residues of chemicals used in controlling pests. Also, given that several types of pests are involved in infestations, policy should be directed toward developing multiple-resistant pulse varieties that simultaneously control many pests.

As discussed, the availability of improved-quality seeds is one of the most important drivers for increasing pulse production. However, it is not only the development of technology but the delivery systems that have been an issue in the case of pulse technology. One way forward may be to link the formal and informal seed sectors for sustainable seed delivery to farmers, which would also motivate small and medium-size seed companies to enter into pulse seed production. Importantly, the private sector has been missing from the research and development in pulses, which is mainly driven by the public sector, including the government and such international organizations as ICRISAT and ICARDA. This is in sharp contrast with crops like maize and pearl millet.

To change this situation, best practices should be drawn from cases where the private sector has delivered in terms of seed development and then applied to pulses. The positive impact of seed policy reforms on private investment in agriculture has been studied and summarized in Kolady, Spielman, and

Cavalieri (2012), who document that although public research organizations and state seed corporations still play an important role in India, the private-sector seed companies have become important. State seed companies are now confined to distributing certified seeds in pulses and other low-value crops such as wheat and rice, while the private sector has made sizable inroads in the higher-value segment of the seed market, which includes hybrids of crops like maize and pearl millet (Pray, Ramaswami, and Kelley 2001).

References

AICRP (All India Coordinated Research Project) on MULLaRP. 2013. *Annual Report 2012/13.* Kanpur, India: Indian Council of Agricultural Research (ICAR).

Ali, M. 2004. "Role of Pulse in Crop diversification." In *Pulses in New Prospective,* 245–254. Kanpur, India: Indian Society of Pulses Research and Development, Indian Institute of Pulses Research (IIPR).

ASTI (Agriculture Science and Technology Indicators). 2014–2015. "India." Led by International Food Policy Research Institute (IFPRI). Accessed June 2016. https://www.asti.cgiar.org/india.

AVRDC. 1999. AVRDC Report 1998. Asian Vegetable Research and Development Center, Shanhua, Tainan, Taiwan. vii + 148 pp.

Bandillo, N., C. Raghavan, P. A. Muyco, M. A. L. Sevilla, I. T. Lobina, C. J. Dilla-Ermita., C.-W. Tung, S. McCouch, M. Thomson, R. Mauleon, R. K.Singh, G. Gregorio, E. Redoña, and H. Leung. 2013. "Multi-Parent Advanced Generation Inter-Cross (MAGIC) Populations." In *Rice: Progress and Potential for Genetics Research and Breeding,* 6–11.

Bantilan, C., C. D. Kumara, P. M. Gaur, M. D. Shyam, and D. Jeff. 2014. *Short-Duration Chickpea Technology: Enabling Legumes Revolution in Andhra Pradesh, India. 2014.* Research Report 23. Patancheru, India: International Crops Research Institute for the Semi-Arid Tropics (ICRISAT).

Bantilan, M. C. S., and P. K. Joshi. 1996. *Returns to Research and Diffusion Investments on Wilt Resistance in Pigeonpea.* Impact Series 1. Patancheru, India: ICRISAT.

Bantilan, M. C. S., and D. Parthasarathy. 1999. *Efficiency and Sustainability Gains from Adoption of Short Duration Pigeonpea in Non-Legume Based Cropping Systems.* Impact Series 6. Patancheru, India: ICRISAT.

Cavanagh C., M. Morell, I. Mackay, and W. Powell. 2008. "From Mutations to MAGIC: Resources for Gene Discovery, Validation, and Delivery in Crop Plants." *Current Opinion in Plant Biology* 11: 215–221.

Gowda, C. L. L., S. Srinivasan, P. M. Gaur, and K. B. Saxena. 2013. "Enhancing the Productivity and Production of Pulses in India." In *Climate Change and Sustainable Food Security*, edited by P. K. Shetty, S. Ayyappan, and M. S. Swaminatha, 145–159. Bangalore: National Institute of Advanced Studies; New Delhi: Indian Council of Agricultural Research.

ICRISAT (International Crops Research Institute for the Semi-Arid Tropics). 2016. "First Machine Harvestable Chickpea Variety Boon for Andhra Pradesh Farmers." Accessed March 2016. www.icrisat.org/first-machine-harvestable-chickpea-variety-boon-for-andhra-pradesh-farmers/.

India, Ministry of Agriculture. 2016. National Initiative for Information on Quality Seeds. Seednet India Portal. Accessed August 2016. http://seednet.gov.in/PDFFILES/SRR-13.pdf.

Joshi P. K., M. Asokan, and M. C. S. Bantilan. 1999. "Silent Chickpea Revolution in Nontraditional Areas—Some Experience from Andhra Pradesh." *Indian Journal of Agricultural Economics* 54 (4, October–December): 115–129.

Kolady, D., D. J. Spielman, and A. J. Cavalieri. 2012. "Intellectual Property Rights, Private Investment in Research, and Productivity Growth in Indian Agriculture: A Review of Evidence and Options." *Journal of Agricultural Economics* 63 (2): 361–384.

Materne, M., and A. A. Reddy. 2007. "Commercial Cultivation and Profitability." In *The Lentil: An Ancient Crop for Modern Times*, edited by S. S. Yadav, D. L. Mc Neil, and P. C. Stevenson, 73–186. In *Operational Guidelines for Rice Fallow Areas*, edited by S. S. Yadav, D. McNeil, and P. C. Stevenson, 173–186. Rotterdam, Netherlands: Springer. Ministry of Agriculture. 2016. Accessed September 2016. Available at http://rkvy.nic.in/static/download/pdf/Rice_Fellow_Guideline.pdf.

Mula, R. P., M. G. Mula, R. S. Gopalan, S. Das, R. V. Kumar, and K. B. Saxena. 2014. Mid-Term Impact Assessment Study—Introduction and Expansion of Improved Pigeonpea (Arhar) Production Technology in Rainfed Upland Ecosystems of Odisha. Project Report. Patancheru, India: ICRISAT.

NAAS (National Academy of Agricultural Sciences). 2013. "Improving Productivity of Rice Fallows." Policy Paper 64. NAAS, New Delhi, India.

Niranjan, S. K. D. F. F., M. C. S., Bantilan, P. K. Joshi, and K. B. Saxena. 1998. "Evaluating Hybrid Technology for Pigeonpea." In *Assessing Joint Research Impacts: Proceedings of an International Workshop on Joint Impact Assessment of NARS/ICRISAT Technologies for the Semi-Arid Tropics*, edited by M. C. S. Bantilan and P. K. Joshi, 231–240. December 2–4, 1996. Patancheru, India: ICRISAT.

Pray, C. E., B. Ramaswami, and T. Kelley. 2001. "The Impact of Economic Reforms on R&D by the Indian Seed Industry." *Food Policy* 26 (6): 587–598.

Rubyogo, J. C., L. Sperling, and T. Assefa. 2007. "A New Approach for Facilitating Farmers Access to Bean Seed." *LEISA Magazine* 23 (2): 27–29.

Ryan, J. G. 1981. "Estimation of the Economic Value of Production Losses due to Diseases of Pigeonpeas." Appendix V. In *International Survey of Pigeonpea Diseases*. Departmental progress report 12. Pulse pathology. Patancheru, India: ICRISAT.

Sarkar, A., S. K. Mishra, N. K. Gautam, and S. K. Sharma. 2007. "Widening the Genetic Base of Lentil in South Asia through the Use of Exotics." In *Proceedings, National Symposium on Legumes for Ecological Sustainability: Emerging Challenges and Opportunities*, 378–391. Kanpur, India: Indian Institute of Pulses Research.

Saxena, K. B. 2009. "Evolution of Hybrid Breeding Technology in Pigeon Pea." In *Milestones in Food Legume Research*, edited by M. Ali and S. Kumar, 82—114. Kanpur, India: Indian Institute of Pulses Research.

Saxena, K. B., R. V. Kumar, N. Srivastava, and B. Shiying. 2005. "A Cytoplasmic-Nuclear Male Sterility System Derived from a Cross between *Cajanus cajanifolius* and *Cajanus cajan*." *Euphytica* 145: 291–296.

Saxena, K. B., and N. Nadarajan. 2010. "Prospects of Pigeonpea Hybrids in Indian Agriculture." *Electronic Journal of Plant Breeding* 1 (4): 1107–1117.

Saxena, K. B., and R. V. Kumar. 2013. "Pigeonpea." In *Hybrid Seed Production in Field Crops*, 213–240. New Delhi-12: Kalyani Publishers.

Shiyani, R. L., P. K. Joshi, M. Asokan, and M. C. S. Bantilan. 2002. "Adoption of Improved Chickpea Varieties: KRIBHCO Experience in Tribal Region of Gujarat, India." *Agricultural Economics* 27 (1): 33–39.

Singh, B. B., G. P. Dixit, and P. K. Katiyar. 2010. *Vigna Research in India (25 Years of Research Achievements)*. All India Coordinated Research Project on MULLaRP. Kanpur, India: Indian Council of Agricultural Research.

Singh, N. P., and M. C. Saxena. 2016. "Towards Self-sufficiency of Pulses in India." Base paper. Strategy Workshop. April 7–8. Indian Council of Agricultural Research. Accessed September 2016. www.taas.in/documents/SW_Pulses_Base%20Paper_April2016.docs.

Singh, N. P., and S. Sewak. 2013. *Global Perspective of Chickpea Research*. All India Coordinated Research Project on Chickpea, 18th Annual Group Meeting, August 24–26, 2013. Jawarharlal Nehru Krishi Visvavidyalaya, Jabalpur, India.

Suhasini, K., N. G. Kulkarni, C. Bantilan, C. D. Kumara, G. D. Nageswara Rao, V. Jayalakshmi, A. G. Vijayakumar, D. M. Mannur, Y. Satish, P. M. Gaur, and K. P. C. Rao. 2012. *Chickpea Baseline and Early Adoption Surveys in South Asia Insights from TL-II (Phase-I) Project: Synthesis Report 2013*. Research Report 19. Patancheru, India: ICRISAT.

Upadhyaya, H. D., R. P. S. Pundir, S. L. Dwivedi, and C. L. L. Gowda. 2009. *Mini Core Collections for Efficient Utilization of Plant Genetic Resources in Crop Improvement Programs.* Information Bulletin 78. Patancheru, India: ICRISAT.

Wani, S. P., O. P. Rupela, and K. K. Lee. 1995. "Sustainable Agriculture in the Semi-arid Tropics through Biological Nitrogen Fixation in Grain Legumes." *Plan and Science* 174: 29–49.

Appendix

TABLE 4A.1 Promising varieties of chickpea, by state

State	Variety
Andhra Pradesh	JG 11, KAK 2, JAKI 9218, MNK–1, ICCV 37
Bihar	Gujarat Gram 4, Pant G 186, HK 05–169, Pusa 372
Chhattisgarh	Digvijay, JG 6, JAKI 9218, JG 14, JG 63, IPCK 2002–29, Vaibhav
Gujarat	JG 16. Gujarat Gram 1, Gujarat Junagadh Gram 3, JSC 55 (Raj Vijay Gram 202), JSC 56 (Raj Vijay Gram 203)
Haryana	Haryana Chana–3, Haryana Chana–5, HK–1
Jharkhand	KPG 59, BG 1003, Pant G 114, KWR 108, Pusa 372, HK 05–169
Karnataka	ICCV 37, JAKI 9218, JG 11, MNK–1, Phule G 0517
Madhya Pradesh	JG 130, JG 322, JG 63, JG 16, JG 14, JAKI 9218, JGK 2, JG 315, JGK–1, Vijay, JSC 55 (Raj Vijay Gram 202), JSC 56 (Raj Vijay Gram 203), Raj Vijay Kabuli 101, Raj Vijay 201, Phule G 0517, PKV Kabuli 4
Maharashtra	Vijay, Digvijay, JAKI 9218, Vishal, Virat, KAK 2, Phule G 0517, JSC 55 (Raj Vijay Gram 202), JSC 56 (Raj Vijay Gram 203), PKV Kabuli 4
Punjab	GPF 2, L 551
Rajasthan	GNG 1581, RSG 888, Pratap Chana–1, GNG 1488, GNG 1499, GNG 663, GNG 469, RSG 973, RSG 963, CSJD 884
Tamil Nadu	JG 11, Co4
Uttar Pradesh	KPG 59, KGD 1168, KWR 108, HK 05–169, Pusa 372
Uttarakhand	Pant G 186, Pant G 114, DCP 92–3, Pant Kabuli 1
West Bengal	Anuradha, Mahamaya–1, Mahamaya–2

Source: AICRIP Annual Reports (various years).

TABLE 4A.2 Promising varieties of pigeon pea, by state

State	Variety or hybrid
Andhra Pradesh	ICPL 151, IT 6, Maruthi, ICPL 87, ICP 332, Asha, Sarita, Durga (ICPL 84031), Laxmi (ICPL 85063), WRG 27, WRG 53, LRG 30, LRG 38, Lam 41
Bihar	Bahar, Pusa 9, NDA 1, DA 11 (Sharad), UP AS 120, BirsaArhar 1
Gujarat	BDN 2, C 11, TT6, TIS-IS, ICPL 87, GT 100, Asha» GAUT 001E(Banas)GTH 1, IT 40I,TJT 501,BSMR 853, GT 101
Haryana	Manak, Pusa 84, Pus a 33, Pusa 855, Paras, Pus a 992, PAU 881
Karnataka	ICPL 151, TT6, ICPL 87, TTB 7, Maruthi, Asha, BRG 1, BRG 2, Hy3C, TS 3R,WRP 1
Madhya Pradesh	C 11, TT 6, ICPL 87, JA 4, Asha, JKM 7, MA 3, IT 401, TJT 501, JKM 189
Maharashtra	BDN 2, C 11, TT6, TAT 10, ICPL 87, BSMR 175, Asha, BSMR 736, JKM 7, AKT 8811, TJT 501, IT 401,Vipu1a, BSMR 853, BDN 708
Odisha	UPAS 120, Asha
Punjab	UPAS 120, AL 5 Manak, Pusa 84, Pus a 33, Jagriti (ICPL 151) , Pusa 855,Al 201,Pusa 992, PAU88I, PA 291
Rajasthan	Pus a 992, UPAS 120, Pusa 855, VLA 1
Tamil Nadu	ICPL 151, IT6, ICPL 87, Co 5, Vamban 1, Asha, Co6, CORG 9701, Vamban 2
Uttar Pradesh	UPAS 120, Bahar, NDA 1, NDA 2, Amar, Azad, MA 6,MAL 13, Pusa 9, NDA 3
West Bengal	Bahar, WB 20,Pusa 9, NDA 2, NDA 3

Source: AICRIP Annual Reports (various years).

TABLE 4A.3 Promising varieties of green gram for different growing seasons, by state

State	Growing season	Variety
Andhra Pradesh	Kharif	PKV AKM 4, IPM 02–14, COGG 912, OUM 11–5, Warangal–2, LGG 407, LGG 450, Madhira 295
	Rabi	Pusa 9072, LGG 460, TM 96–2, WGG–2
Assam	Kharif	SG 1 (Pratap), Pant moong 2, Pant Moong 4, Narendra moong 1, IPM 2–3
	Spring/ summer	PDM 139, Pusha Vishal, Meha, Pant moong 5, TMB 37, HUM–16, HUM 12
Bihar and Jharkhand	Kharif	Pant moong 2, Pant Moong 4, Narendra moong 1, Sunaina, PD-M139,MH2–15, HUM–1, IPM 2–3
	Spring/ summer	PDM 139, Pusha Vishal, Meha, Pant moong 5, TMB 37, HUM–16, HUM 12
Delhi	Kharif	IPM 2–3, Pant Moong 3, ML 337, MUM 2, Ganga 8, MH 02–15
Gujarat	Kharif	PKV AKM 4, BM 4, Gujarat Moong 3, Pant moong 2, PIMS 4 (Sabar-mati), GujratMoong 2, GujratMoong 4
	Spring/ summer	Gujarat moong 2, PDM 139
Haryana	Kharif	MUM 2, Pusa Vishal, Ganga 8, MH 2–15, IPM 2–3, Muskan
	Spring/ summer	Pusa Vishal, SML 668, Pant Mung–5
Himachal Pradesh and Jammu and Kashmir	Kharif	Pant Moong 2, Pant Moong 6, KM 2241, Shalimar moong 1, Pusa 0672
Karnataka	Kharif	IPM 02–14, PKV AKM 4, COGG 912, HUM 1, China Moong, KKM 3
Madhya Pradesh and Chhattisgarh	Kharif	Pant Moong 3, ML 337, BM 4, JM 721, Jawahar 45, HUM–1, Meha, TJM 3
	Spring/ summer	HUM 1, Pusa 9531, PDM 139, Meha
Maharashtra	Kharif	PKV AKM 4, Kopergaon, ML 131, BM 4, Phule M 2, TARM 1, TARM 18, TARM 2, BM 200–1, HUM 1
Odisha	Kharif	OUM 11–5, COGG 912, PKV AKM 4, TARM 1, PDM 139
	Rabi	Pusa 9072, Sujata (Hyb 2–4), TARM–1, OBGG–52, LGG–460, PDM 139
Punjab	Kharif	MUM 2, ML 613, Ganga 8, MH 2–15, IPM 2–3
	Spring/ summer	Pusha Vishal, Pant moong 2, SML 668, Pant Mung–5
Rajasthan	Kharif	Ganga 8, RMG 268, MUM 2, SML 668, RM 492, IPM 2–3, MH 2–15
	Spring/ summer	RMG 268, SML 668, PDM–139, Meha
Tamil Nadu	Kharif	Paiyur 1, Vamban 1, ADT 3, CO 5, TM 96–2, COGG 912, OUM 11–5
	Rabi	Pusa 9072, Sujata (Hyb 12–4), ADT–3
Uttar Pradesh and Uttrakhand	Kharif	Pant Moong 2, Pant Moong 3, Narendra Moong 1, Pant Moong 4, Pant Moong 5
	Spring/ summer	PDM 139, Pusha Vishal, Meha, Pant moong 5, TMB 37, HUM–16, HUM 12
West Bengal	Kharif	Narendra Moong 1, Pant Moong 4, Pant Moong 5, MH 2–15, Sreku-mar
	Spring/ summer	PDM 139, Pusha Vishal, Meha, Pant moong 5, TMB 37, HUM–16

Source: AICRIP Annual Reports (various years).

TABLE 4A.4 Promising varieties of black matpe for different growing seasons, by state

State	Growing season	Variety
Andhra Pradesh	Kharif	WBG 26, KU 301 (Shekhar –1), WBU 108, LBG 648, Pant U 31, IPU 2–43, LBG 685, LBG 625, LBG 752, IPU 07–3, VBG 04–008, LU 391
	Rabi	TU 94–2, LBG 611, LBG 20, LBG 402, LBG 623, LBG–709, WBG–26
Assam	Kharif	Pant U 30, WBU 108, IPU 94–1 (Uttara), WBU 108
Bihar and Jharkhand	Kharif	IPU 94–1 (Uttara), BirsaUrd 1, Pant U 30, Pant U 31, WBU 108
	Spring	KU 92–1 (Azad Urd 1), WBU–109, Pant U 31
Gujarat	Kharif	KU 96–3, TPU 4, AKU 4, WBU 108, GU 1
Haryana	Kharif	Mash 338, Pant U 19, KU 300 (Shekhar 2), WBU 108, IPU 94–1 (Uttara)
Himachal Pradesh	Kharif	Pant U 19, Pant U 31, Pant U 40
Karnataka	Kharif	KU 301, WBG 26, WBU 108, LBG 402, LBG Manikya, 1, TU 94–2, LU 391, IPU 07–3, VBG 04–008, IPU 2–43
Madhya Pradesh and Chhattisgarh	Kharif	KU 96–3, TPU 4, JawaharUrd 2, JawaharUrd 3, Khargone 3, Pant U 30
	Spring	Pant U 31
Maharashtra	Kharif	TPU 4, Pant U 30, TAU 1, TAU 2, AKU 4 (Melghat), AKU 15, KU 96–3
Odisha	Kharif	KU 301, WBG 26, WBU 108, Sarla, IPU 2–43
	Spring	TU 94–2, LBG 402, OBG 17, B–3–8–8, Mash 338
Punjab	Kharif	IPU 94–1 (Uttara), WBU 108, Krishna, Mash 414
	Spring	KU 300 (Shekhar – 2), KUG 479
Rajasthan	Kharif	IPU 94–1 (Uttara), WBU 108, Pant U 31, KU 300
	Spring	KU 300, KUG 479
Uttar Pradesh and Uttrakhand	Kharif	IPU 94–1 (Uttara), WBU 108, Narendra Urd 1, Pant U 35, Pant U 31, Pant U 40
	Spring	KU 92–2 (Azad Urd 1), KU 300 (Shekhar 2), Narendra Urd 1, WBU 109, KUG 479
Tamil Nadu	Kharif	ADT 3, ADT 5, Vamban 2, WBU–108, KU 301 (Shekhar 1), Vamban–3, ADT 4,Vamban–4, ADT 5, IPU 07–3, IPU 2–43, VBG 04–008 WBG–26, Vamban–3, TU 94–2, VBN–5, IPU 2–43. KBU 512, Vamban 2
West Bengal	Kharif	IPU 94–1 (Uttara), WBU 108, Pant U 31
	Spring	KU 92–1 (Azad urd 1), WBU 109, Pant U 31

Source: IIPR Annual Reports (various years).

TABLE 4A.5 Promising varieties of lentil, by state or region

State or region	Variety
Assam	HUL 57, WBL 77, KLS 218, Asha (B 77)
Bihar	HUL 57, WBL 77, Arun (PL 77–12)
Delhi	DPL 62 (Sheri), LH 84–8
Gujarat	IPL 81, JL 3
Haryana	DPL 62 (Sheri), IPL 406
Himachal Pradesh	HUL 57, VL 507
Jammu and Kashmir	HUL 57, VL 507, Shalimar Masoor 1
Madhya Pradesh	IPL 81 (Noori), JL 3, IPL 406
Maharashtra	IPL 81 (Noori), JL 3
Northeast Hill Region	HUL 57, DPL 62
Odisha	HUL 57, WBL 77, B 77 (Asha)
Punjab	DPL 62 (Sheri), Pant L 4, LH 84–8, LL 147
Uttar Pradesh	HUL 57, DPL 62 (Sheri), IPL 81 (Noori), Narendra Masoor 1, IPL 406
West Bengal	HUL 57, WBL 77 KLS 218, Ranjan (B 256), Asha (B 77)

Source: IIPR Annual Reports (various years).

Chapter 5

STRUCTURE OF PULSE PROCESSING IN INDIA

Devesh Roy and Raj Chandra

The pulse processing industry has the potential to play a special role in India because of the way pulses are consumed in the country and because of possible backward links that could uplift farmers living in marginalized environments. Pulse processing could also enhance the incomes of farmers and other participants in the value chain for crops that have not been among the best performers for a long time. In this chapter, we analyze the pulse processing sector's growth and the relative roles of the organized and the unorganized sectors of the industry. We also identify the constraints facing pulse processing and suggest a way forward for the sector.

Background

A significant number of mills in India's pulse processing sector remain part of the unorganized and small-scale manufacturing sector that uses traditional technology, while 75 percent of the pulses produced in India are processed in organized sector dal mills (Banerjee and Palke 2010).[1] Although a large quantity of pulses are processed by medium-size industries, a significant amount is still processed in the rural sector without proper machinery. This not only affects the availability of dal in the rural sector due to loss during processing but also results in an inferior quality product that fetches lower prices. According to NAAS (2006), traditionally milled dal fetches 20 percent less in the market than the average quality dal and hence is generally sold in the rural market only. On the one hand, the government has promoted commercial food processing with several initiatives, deregulating and de-licensing it after 1991 reforms (except for alcoholic beverages). On the other hand, several

1 *Dal* is split pulses; it is the product that results after primary processing. The term *unorganized sector* in India refers to unincorporated private enterprises owned by individuals or households engaged in the sale or production of goods and services operated on a proprietary or partnership basis and with fewer than 10 total workers.

policy bottlenecks remain that constrain the sector. The pulse-milling sector, which earlier had been reserved for the small-scale sector, was "de-reserved" in the late 1990s. As such, no license or permission is now required for setting up a pulse mill, apart from permission from the Departments of Health, Industries, and Pollution Control Board.

Regarding the impact and opportunities of global trade, policy has taken several turns over the past 25 years. The excise duty on food-processing items was removed in 1991, then reimposed in 1997, only to be removed again in 2001. In the food-processing industry, including pulses, the government gives automatic approval to foreign investment of up to 100 percent equity, except in a few cases; 100 percent export-oriented units (EOUs) are permitted to import raw material and capital goods free of duty. Moreover, in agro-based industries, EOUs are allowed to sell up to 50 percent of their products in the domestic tariff area (Dev and Rao 2004). The concept of food parks and agri-export zones (AEZ) has been initiated along with several incentive schemes (Dev and Rao 2004). Despite all of this, the value-addition in foods in India remains quite low today, measured at 7 percent in India as compared with 23 percent in China.

Several policy-driven factors still inhibit India's food-processing sector, government initiatives discussed above notwithstanding. According to the Department of Scientific and Industrial Research (DSIR), while the incidence of tariffs and indirect taxes has been reduced over the years, the tax structure for a range of processed foods (including pulses) is not uniform and not conducive to the processing sector. There is a different tax structure for branded and nonbranded food items, for example, and since branded items attract higher sales tax, they are costlier. Such imposed costs have adversely affected the processed food sector. Processed foods in India are costlier than fresh foods, unlike the situation in other countries. This stems from a series of taxes and duties applied in India. Most countries in the world do not levy taxes or duties on processed food products, seeking instead to promote value-addition in the food sector (India, DSIR 2007). Under the value-added tax (VAT), which is levied by state governments, most processed food products are taxed at varying rates of 1 percent, 4 percent, and 13 percent. Apart from VAT, other taxes (such as entry tax and octroi) are levied on food products. Also, the packing material attracts a high excise duty of 12 percent, further raising the costs of processed foods. Even the customs duty on packaging materials continues to be high. Consequently, the net tax effect ranges from 21 percent to 23 percent on various food items (Dev and Rao 2004).

Study Objectives and Data Sources

Since the pulse processing industry in India has remained in the preliminary stage of development for some time, with a significant number of unorganized-sector mills, in this study, we first look at specific attributes of pulse processing in both the organized and unorganized sectors. We then discuss the evidence for a changing relative share of the organized sector of pulse processing and examine growth rates over time. Next, we analyze some of the variables that may be undergoing a restructuring in the processing industry, such as scale of operation and capital-to-labor ratio. We conduct a regression analysis to identify the determinants of firms' scale and productivity in the industry, which enables a glimpse into possible new growth areas. Toward the end of the chapter, we briefly look at the supply chain. The final section summarizes the findings and makes concluding observations along with some policy implications.

The analysis uses two secondary sources of data on the pulse processing sector: the National Sample Survey (NSS) for data on manufacturing units in the unorganized sector, and the Annual Survey of Industries (ASI), collected by the Central Statistical Organization (CSO), for data on the organized (incorporated) segment. The enterprises in the organized sector are those registered under the 1956 Company Act of India. One limitation is that the data from these two available sources are not disaggregated by type of pulse; instead, the data are combined for the pulse processing sector as a whole. The profile of pulse processing mills in the organized sector is analyzed using ASI unit-level data. For both NSS and ASI, the prescribed sample weights were used to arrive at the estimated population figures. For comparability across time, the monetary variables were deflated using the Wholesale Price Index (WPI) with base 2004–2005. WPI better monitors price movements that reflect demand and supply in industry, manufacturing, and construction sectors and is used by the government in measuring inflation. Greater use of the consumer price index (with revisions) has become common only recently. Note that neither of these sources provides data for processing broken down by type of pulse; instead, the data cover aggregate pulse processing, which at a higher level is part of grain milling.

Key Findings

The evidence examined in this chapter shows that India has experienced an unambiguous scaling up in output per factory and in capacity—that is, in fixed capital per factory. This suggests that a production reconfiguration is going on involving new technologies and products at the factory level. Based on the data, overall the scale of individual pulse mills and the employment they generate remain quite small, not only in the unorganized but also in the organized sector. Moreover, the capital-to-labor ratios are lower in pulse processing than they are in most other food-processing sectors. For example, in 2006 the capital-to-labor ratio in sugar processing was more than 10 times greater than that in pulse processing (Bhavani, Gulati, and Roy 2006). Both as an engine of growth and for the upgrading of technology, it is the organized sector that is likely to be the prime driver, with data showing that in relative terms the organized sector has expanded in recent years. Measured by the number of mills, the unorganized sector remains comparatively large, but in terms of output there is a shift increasingly favoring the organized sector. Data also show that some states and regions are well ahead of others regarding the relative importance of the organized sector. The pattern seems to show that in poorer and more agrarian states, nearly all the processing is very small-scale and unorganized, while the organized and large-scale factories are clustered in urban areas. As a result, the most efficient pulse processing occurs in places far removed from most of the pulse-growing areas.

The analysis assesses factory-level restructuring by examining changes in structural characteristics, such as scale of operation, technological change, and productivity.[2] When we assess the factors that account for mills' productivity, we find strong evidence of state-specific factors playing a role. Moreover, there is mixed evidence for some agglomeration effects, such that where there is more processing activity (in terms of output but not in terms of the number of mills) in a neighborhood, the average processing mill shows better performance. In terms of state-specific effects treating the leading state, Madhya Pradesh, as a benchmark, several states have significantly lower mill productivity. Finally, we find that years of operation have a significant bearing on labor productivity, indicating that "learning by doing" could be playing a role.

The Mechanics of Pulse Processing

The processing of dal is the second-largest food-processing industry in India after rice milling and flour milling. The essential work of a dal mill follows

2 At a firm level, restructuring could be examined through mergers and acquisitions, but we do not have the data needed to analyze the issue that way, hence we focus on the industry- and factory-level characteristics.

three stages: dehulling the pulses, splitting them, and grinding into flour. According to the India Pulse Growers Association (IPGA), 75 percent of pulses produced in India are processed. Therefore, postharvest technology—whether it is traditional or advanced—plays an important role in the per capita availability of pulses. Pulse mills vary in size, from cottage industries to sophisticated factories with pneumatic conveyors. Nevertheless, most of the pulse processing industry is small scale, comprising thousands of dal mills distributed throughout the country whose daily capacity is small, ranging from 0.5 ton per day to 10 tons per day. Measured in terms of both the number of mills and employment, a large part of the pulse processing is in the unorganized sector; these mills use conventional technology with locally fabricated machinery. Geographically, pulse mills (of all sizes) are concentrated in the producing areas such as Indore (Madhya Pradesh), Jalgaon and Akola (Maharashtra), and in some big cities such as Chennai, Delhi, Hyderabad, Kolkata, and Mumbai. It is estimated that there are about 10,000 pulse mills in India, mostly in the private sector. On average they operate for 200 to 250 days per year.

Primary versus Secondary Processing

For pulses, *primary processing* consists of dehusking the grain, splitting it into dal, and grinding it into flour. *Secondary processing* refers to treatments that convert the dal and flour into acceptable, edible products (ICRISAT 1991). The processing follows these steps: The pulses are cleaned and foreign matter, such as stones and mud, is removed. Next, the surfaces of the pulses are scratched, which improves absorption when they are next soaked in a mixture of water and vegetable oil (known as *dampening*). After they are dried (known as *tempering*), they are dehusked through grinding. Once the outer layer is removed, the pulses are split in half. These primary processes do not greatly influence the nutrient composition or acceptability of the pulses. Some additional practices like polishing with marble or leather polish, though not part of standard primary processing, can have health hazards or nutrition-depletion effects (see the discussion about polishing of pulses in this context in Chapter 7).

Dals that are split—as in the case of pigeon pea, black matpe, green gram, and lentil—are more difficult to dehusk, so they require repeated operations by dehusking rollers. The soaking and drying mentioned above are repeated to loosen portions of husk that may be sticking even after repeated rolling. Linseed oil is often used to impart shine to the milled dal, which is appealing to consumers. Some pulses—mostly chickpea, black matpe, and green gram—are milled to make flour (*besan*) through grinding. To give a better finish, some processors polish the dal. There is a belief that polishing leads to nutrient loss in pulses. Unpolished pulses are sold by one of the processors, Tata i-Shakti, as a differentiated health product because of this attribute.

Secondary processing varies widely, depending on the consumers being targeted. One method involves soaking seeds in an alkaline solution to increase the water uptake of dal during cooking and decrease cooking time while increasing the dispersion of solids (Chavan et al. 1983). Unlike primary processing, elements of secondary processing do significantly influence the nutritional quality of pulse products. The processing time used, the temperature, and the moisture level are three important factors in this regard. Moist heat methods are considered better than dry heat methods (Geervani and Theophilus 1980), and processing for longer than 10 minutes above 120°C is reported to cause considerable damage to proteins (Rama Rao 1974). This is quite important given that pulses are valued as a provider of protein. The secondary processing may involve a variety of dry or moist heating: roasting, boiling, steaming, or frying. Products using chickpea flour often require converting the flour into batter, which is then fried or else fermented and steamed. By controlling the proportion of water to flour in a batter, fried products of varied texture can be prepared (ICRISAT 1991). "Puffing" chickpeas, a secondary process, is a cottage industry in India.

Processing Efficiency

According to the IPGA, the output of the mills depends closely on the availability of raw material, capital, and energy, as well as the capacity of each mill and the number of working days it operates. The different technologies involved in pulse processing have different levels of sophistication, depending on the properties of the grain and the efficiencies needed. Technology is likely the main differentiator between organized and unorganized pulse processing mills. To increase their use and consumption, in addition to their own processed variants, pulses are used as ingredients in other food items. Examples include breads, condiments, snacks, flours, gels, noodles, pasta, other baked goods, protein analogs, and snacks. The functionality of pulses as ingredients varies by the type of pulse milled and the type of milling procedure employed. The functional properties relevant to their use as ingredients include such attributes as water absorption capacity, oil absorption capacity, taste, texture, cooking time, and color, among others.

The efficiency of pulse processing varies greatly according to the methods used. According to Parpia (1973), domestic small-scale milling processes give yields on the order of 75 percent from chickpea and 68 percent from pigeon pea, whereas improved milling technologies give yields of more than 80 percent, with a theoretical maximum of 89 percent. According to Banerjee and Palke (2010), to minimize losses in processing the dal, industry should maximize the use of improved dal mills, which are highly versatile, technology savvy, and more energy efficient than the traditional mills. These new and improved dal mills have a dehusking efficiency of approximately 95 percent, while their split-pulse yields run between 80 percent and 85 percent, largely depending on the variety of the pulse and the conditioning of the pulse grain. Many agricultural universities and institutions recognized by the Indian Council of Agricultural Research (ICAR) have played a large role in developing such improved dal mills.[3] Today, new options have become available in food processing, including technologies for processing whole pulses, techniques for fractionating pulses into ingredients that preserve their functional and nutritional properties, and other potential applications to incorporate pulses into new food products.

An Economic Analysis of the Organized and Unorganized Sectors of the Pulse Processing Sector

As mentioned earlier, a salient feature of the Indian food industry in general, and its pulse processing sector in particular, has been the preponderance of the unorganized sector, which consists of numerous small units. At the same time, the organized sector of the total food-processing industry has been steadily growing for more than two decades, increasing from 64 percent of the total output in 1984–1985 to 81 percent in 2000–2001, and it is expected that it has grown further since (Bhavani, Gulati, and Roy 2006). An important correlate of the increasing formalization of the food-processing sector is a greater incidence of contractual relationships between farmers and processors (Gulati, Joshi, and Landes 2008). The policy reforms made in India in the early 1990s, such as de-licensing and dereservation for small-scale firms, and those made in

3 Examples of these institutes are Panjabrao Deshmukh Krishi Vidyapeeth (PDKV) in Akola, Central Food Technology Research Institute (CFTRI) in Mysore, and Central Institute of Agricultural Engineering (CIAE) in Bhopal.

the 2000s, such as the launch of mini food parks, in combination with consumers' changing incomes and tastes, were expected to encourage the industry's organized sector, including the pulse processing segment.[4] However, although organized pulse processing dominates in output and has increased its fixed assets over time, comparative data show that the unorganized sector remains dominant in generating employment, although the converse is true for its output and share in sales. In fact, the share in output of the unorganized sector has declined from an already low 26 percent in 2001 to just 23 percent in 2010 (Table 5.1).

Employment

Table 5.1 presents the comparative picture for the organized and unorganized pulse processing sectors in India. Measured by the number of mills, nearly 85 percent of pulse processing is in the unorganized sector. Measured by employment, the unorganized sector also has a large total share of employment, at 70 percent (although this last number might be a bit overestimated, because employment in the unorganized sector includes part-time workers). The large share in total employment is partly due to the technology used, but it could also be policy-induced (for example, relating to labor market regulations). It is also a function of the number of unorganized mills. Hence, even though a typical mill in the unorganized sector employs fewer people than one in the organized sector because the number of unorganized mills is comparatively large, the total employment in that segment is higher. In addition, the traditional manual technologies that most of the unorganized units operate with may require them to employ more workers. In contrast, organized mills

4 Under compulsory licensing under the Industries (Development and Regulation) Act of 1951, industries were put under compulsory licensing on account of such factors as environmental, safety, and strategic considerations. Furthermore, the industrial policy of small-scale reservation forms a significant aspect of India's industrial policy. Dereservation of such items is undertaken by the government at periodic intervals. All undertakings other than the small-scale industrial undertakings engaged in the manufacture of items reserved for manufacture in the small-scale sector are required to obtain an industrial license and undertake an export obligation of 50 percent of the annual production. The exceptions to licensing are for undertakings operating under 100 percent Export Oriented Undertakings Scheme, the Export Processing Zone (EPZ), or the Special Economic Zone Schemes (SEZS). The government of India, in order to promote food processing, has chosen two schemes: namely, mega food parks and mini food parks. The former is based in larger areas between 50 to 100 acres, depending on the business plan, and have central processing centers combined with primary processing centers and collection centers. Mini food parks are with much smaller areas allocated (30 acres) and need not have primary processing centers or collection centers. The extent of subsidy (in the form of financial assistance) is also lower for the mini parks.

TABLE 5.1 Characteristics of the unorganized and organized pulse processing sectors over time, 2001–2010

	2001	2003	2005		2010	
Characteristics	Unorganized	Organized	Unorganized	Organized	Unorganized	Organized
Number of mills	7,897	900	8,496	1,330	8,034	1,517
Total number of workers	23,830	10,773	27,812	13,688	29,058	12,642
Number of men	17,309	7,321	24,427	9,186	24,785	10,027
Number of women	6,521	3,452	3,385	4,502	4,273	2,615
Percentage of men in total workforce	74	68	88	67	85	79
Percentage of women in total workforce	26	32	12	33	16	21
Average number of worker per mill	3.01	11.97	3.42	10.29	3.61	8.33
Fixed capital per mill (Rs millions)	0.05	1.44	0.09	0.88	0.33	3.37
Gross value-added per worker (Rs millions)	0.37	0.17	0.37	0.15	0.09	0.98
Gross value added per mill (Rs millions)	1.12	2.01	1.22	1.58	0.31	8.13
Capital-to-labor ratio	0.018	0.12	0.029	0.09	0.09	0.40
Total output (Rs millions)	13,760	38,731	43,569	77,010	81,480	266,152
Share in total output (%)	26.21[a]	73.78	36.13	63.86	23.41	76.48
Share in gross value-added per mill (%)	35.78	64.21	43.57	56.42	3.67	96.32
Share in gross value-added per worker (%)	68.51	31.48	71.15	28.84	8.41	91.58

Source: Authors' calculations based on data from Annual Survey of Industries and National Sample Survey Organization (NSSO).

Note: [a] The percentage share of the organized sector and the unorganized sector has been calculated by combining the values for 2001 and 2003 to match the data for the organized sector and the unorganized sector to the nearest time period available.

operate with technologies that are more capital- and skill-intensive and therefore less labor-intensive.[5]

This is reflected in much lower rates of capital per worker as well as capital employed per mill in the unorganized sector. Overall, pulse processing does not seem to be a big employment generator, with employment values of just 3.6 workers per mill and 8.33 workers per mill in the unorganized and organized sectors, respectively. In terms of gender composition of employment, the organized sector actually has a much greater share of women in the workforce. Moreover, the scale is much higher in the organized mills as captured in terms of capital per mill and output per mill.

Capital

Table 5.2 presents the dynamics in the organized pulse processing sector over a recent nine-year period. Over that period there was a 70 percent rise (that is, a 6.1 percent compound annual growth rate) in the number of mills in the organized sector, and this was accompanied by substantial capital deepening as reflected in the growth in fixed capital. The growth in fixed capital, however, was lower than growth in working capital. The scaling up of the organized sector is reflected in the amount of inputs used. Between 2002 and 2012 the amount of fuel used in organized pulse processing mills went up, from 549 million rupees to 4,733 million rupees, a 700 percent increase over 10 years.

Fixed capital includes both the plant and the machinery, among other things, and thus captures technology to the extent that it is embodied in the machinery. If the additional machinery is the same as that of the existing machinery, it amounts to capacity addition. If the additional machinery is superior to the existing machinery, it is treated as technological progress (Bhavani, Gulati, and Roy 2006). Growth of fixed capital reflects capacity additions and/or technological progress and thus a rise in the potential scale of operation. Capital deepening can roughly represent technological progress, since new technologies are more capital-dependent and less labor-dependent. The variations in capital deepening evident across mills might be a function of an initial capital intensity that could vary by types of pulses as well as the final market. At the same time, on the supply side, factors like investment measures and competition and policy regulations could determine the outcomes

5 New technologies vis-à-vis traditional manual technologies enable the production of hygienic and standardized products and can thus improve the level of sales.

TABLE 5.2 Organized sector pulse processing over time, 2003–2012 (Rs millions at constant prices for values)

Sl No.	Item	Unit	2003	2007	2012
1	Number of factories	Numbers	900	1,315	1,535
2	Fixed capital	Rs millions	1,294	2,184	13,276
3	Working capital	Rs millions	3,666	12,643	50,300
4	Outstanding loans	Rs millions	3,142	11,287	37,437
5	Man days workers	Thousands	4,050	4,973	7,467
6	Number of workers	Numbers	10,744	12,939	21,338
15	Total inputs	Rs millions	36,949	93,216	478,268
17	Value of output	Rs millions	38,731	97,124	540,937

Source: Authors' calculations based on data from Annual Survey of Industries.

for pulse processing mills. There has been a substantial rise in the fixed investment in the organized pulse processing sector over time, from about 988 million rupees (in constant terms) in 2002 to 13,276 million rupees in 2012. This is due to a significant jump in the capacity—that is, the fixed assets per mill. This capacity expansion may be due to pent-up demand for most products that ignited market expectations for the industry, together with liberalization of investment restrictions on the supply side.

On the employment side, it is possible that the quality of employment in the organized sector, in terms of labor productivity, would be much higher than that found in the unorganized sector. If so, it could be because labor in the organized sector has significantly more capital to work with.

Geographic Distribution

There is significant regional concentration in pulse processing by sector type across states (Figure 5.1). Four states account for the majority of organized pulse processing. Strikingly, some big states (like Rajasthan and Uttar Pradesh) have very few organized sector mills. At the same time, Gujarat, although not a large producer of pulses, is comparatively industrialized and has a large number of organized mills. Highly agricultural states (like Punjab and Uttar Pradesh) do not contain a large number of mills in the organized sector.

Figure 5.2 presents the distribution of mills in the unorganized sector across states. It is notable that not only does a big state like Uttar Pradesh have a large number of unorganized sector mills (nearly 1,500), but a small state like

FIGURE 5.1 Distribution of mills in the organized pulse processing sector, 2010–2011

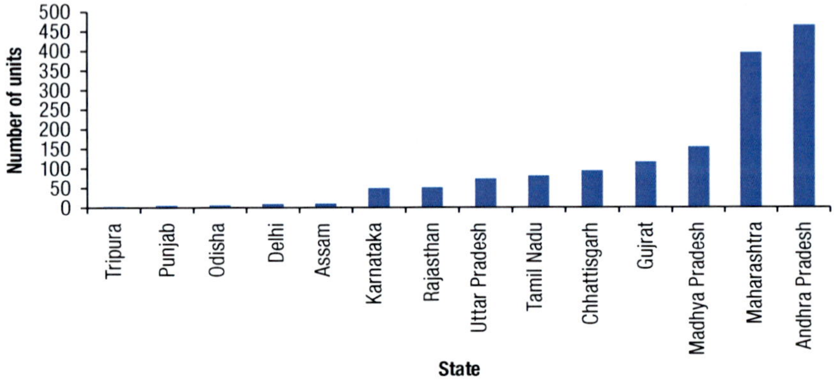

Source: Data from Annual Survey of Industries.

FIGURE 5.2 Distribution of mills in the unorganized pulse processing sector, 2010– 2011

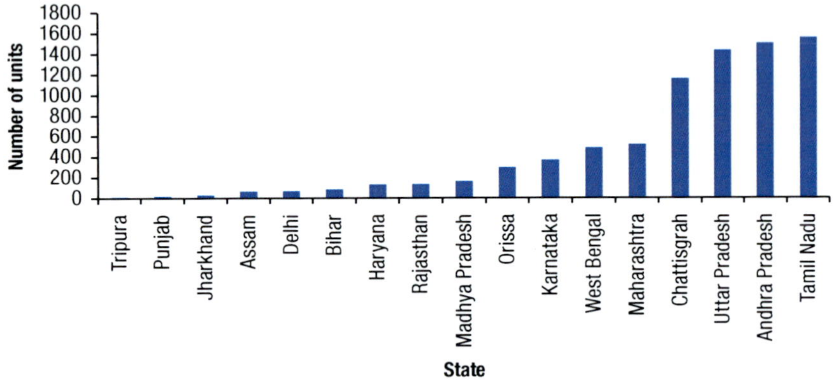

Source: Data from Annual Survey of Industries.

Chhattisgarh (erstwhile part of Madhya Pradesh), which is a big pulse producer but is not industrialized, also has a large number of mills (1,161) in the unorganized sector compared to fewer than 100 mills in the organized sector. Also striking are the cases of undeveloped states, like Bihar (not shown) and Odisha (a state where pulse production has experienced a turnaround in recent times), where the organized processing sector has almost no presence.

Figure 5.3 presents the employment shares of states in the organized and unorganized pulse processing sectors. Nearly 75 percent of employment in the organized sector is concentrated in the four states of Andhra Pradesh,

FIGURE 5.3 Employment shares across states in the organized and unorganized pulse processing sectors in 2010–2011 (%)

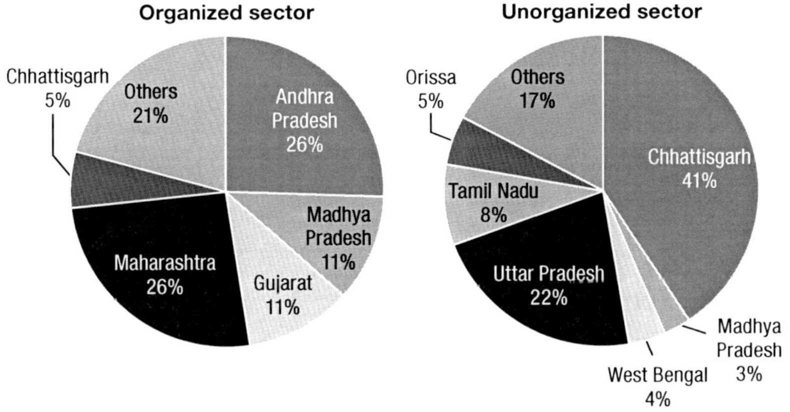

Source: Data from Annual Survey of Industries and National Sample Survey Organization (NSSO).

FIGURE 5.4 Share of fixed capital across states in the organized and unorganized pulse processing sectors (%)

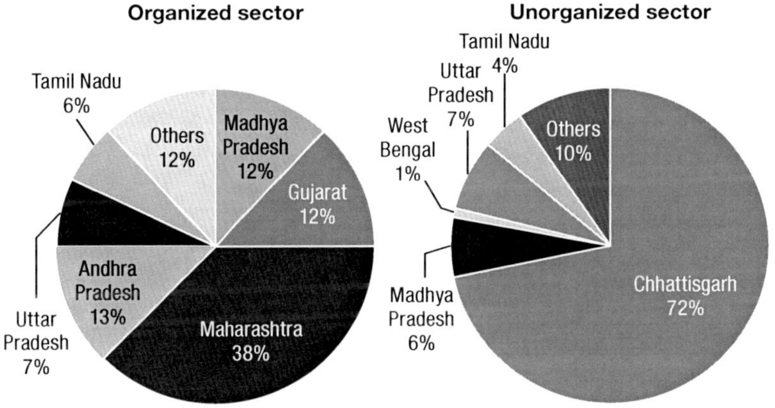

Source: Data from Annual Survey of Industries and National Sample Survey Organization (NSSO).

Gujarat, Madhya Pradesh, and Maharashtra. In the unorganized sector, nearly 70 percent of employment is located in three states: Chhattisgarh, Tamil Nadu, and Uttar Pradesh. These findings are unsurprising given the similar distribution of mills across states in Figures 5.1 and 5.2, respectively.

Figure 5.4 plots the distribution of mills in terms of their fixed capital across states. The pattern mirrors that of employment share, with fixed-capital shares concentrated in a few states. Although pulse processing is spread across the country, these figures show concentration by type of industry. Alternatively, they imply that barring a few states, in a majority of states the processing remains dominated by the unorganized sector (in terms of the number of mills) with low technological intensity. The four dominant states in organized pulse processing account for nearly 80 percent of output in this sector.

Productivity Growth: Shifts in Capital and Labor

The growth in the organized sector of pulse processing could be happening at both the intensive and the extensive margins. New organized mills could be established but, equally, old (unorganized) mills could upgrade in terms of technology. There could be a drop in the number of unorganized mills at the same time as the number of organized mills is increasing, a Schumpeterian example of creative destruction. Over time, the states that have witnessed large drops in the number of unorganized pulse mills are Andhra Pradesh, Bihar, Madhya Pradesh, Rajasthan, and West Bengal. The growth in organized pulse processing has mainly been in urban units. The rural enterprises, meanwhile, remain dominated by unorganized mills.

The share of capital per worker is higher in the leading states for organized pulse processing, so the spatial patterns are also reflected in labor productivity. A few striking facts stand out regarding labor productvity. First, before 2009 output per worker was uniformly quite low across states, and after 2009 it switched to a significantly rising trajectory. It is possible the food price crisis of 2008 created pressure on the processing sector to become more efficient. At the same time, in the organized sector the variance in labor productivity across states also increased after 2009. In essence, capital deepening or technological progress, which was comparatively uniform across states until 2007, became more variable after the food price crisis. Figure 5.5 and Figure 5.6 plot the output per factory and output per worker, respectively, in this sector over time. Meanwhile, in the unorganized sector, output per worker almost doubled over that decade, although it rose from a very low base.

With greater productivity measured as output per worker, the wages and salaries are likely to have been higher in the organized pulse processing sector after 2009 (Figure 5.6).[6] There have been substantial increments in labor productivity in the organized sector, when measured this way, in frontline

6 The *quality* of employment, which determines the wages and hence incomes of employees, is as important as the *quantity* of employment generated.

FIGURE 5.5 Output per mill in the organized pulse processing sector over time, 2001–2012

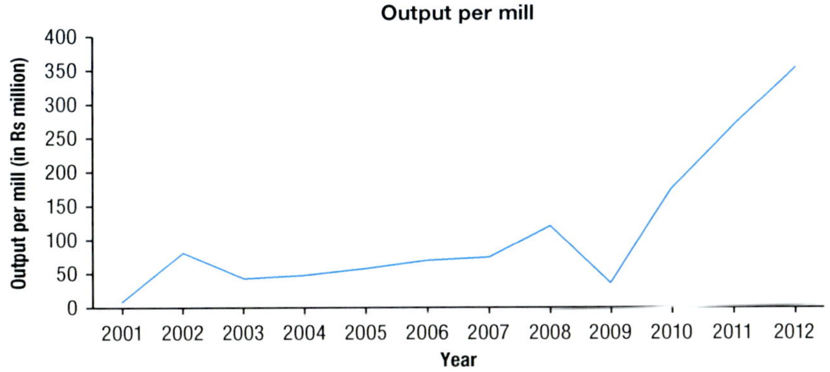

Source: Authors' calculations based on data from Annual Survey of Industries.

FIGURE 5.6 Labor productivity across states over time in organized sector mills, 2002–2012

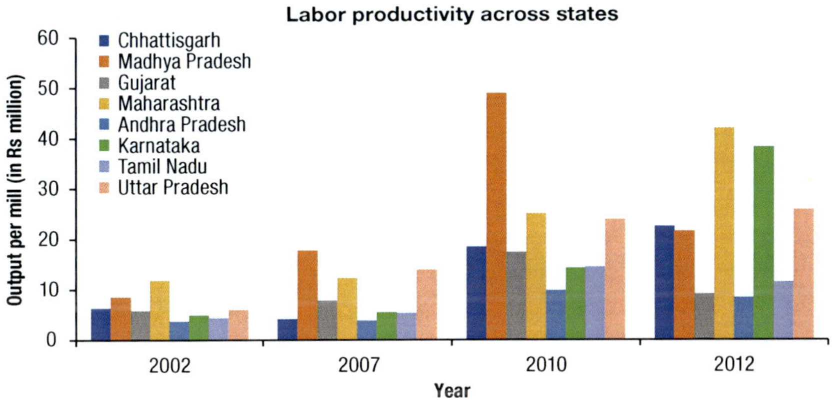

Source: Data from Annual Survey of Industries.

states after 2009. Empirical evidence for the changes in scale of operation of the pulses processing sector indicates that in the nine years following 2001, growth in the value of output per factory was quite modest. An average pulse processing factory in the organized sector produced annual output worth 85 million rupees at the beginning of the period and 300 million rupees at the end. At the factory level, growth in output may be due, on the supply side, to capacity additions or to technological progress, and on the demand side, it may be due to growth in the market. The supply-side factors enhance the capacity of a factory to produce more output, while the demand-side factors

FIGURE 5.7 Fixed capital per worker across states over time in organized sector mills, 2002–2012

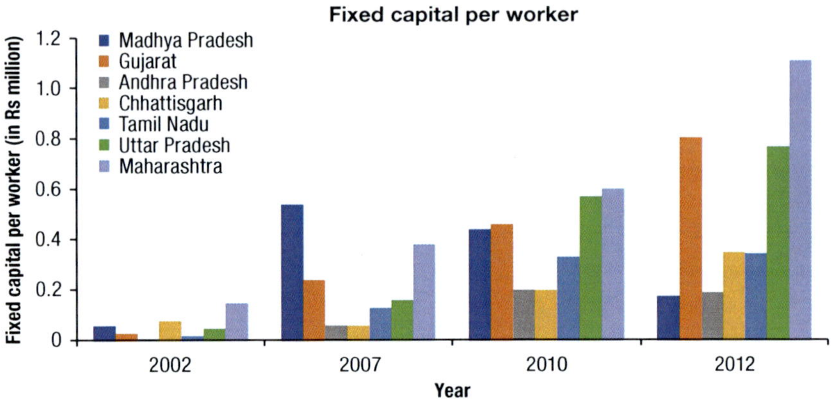

Source: Data from Annual Survey of Industries.

provide incentives to produce more. Consequently, variations in the growth of output per factory across mills could be due to variations in the supply and demand factors as well as to their interactions.

Since we do not have data on market expansions for the mills, we try to capture capacity expansions and technological progress through the growth of fixed capital, capital deepening (the capital-to-labor ratio), and labor productivity (Figure 5.7). Expectations about the market growth and liberalization of investment regulations and the resulting forces of market competition might have prompted the organized-sector firms to bring in new technologies and add to their capacities. If so, that could be reflected in a rise in fixed capital at the factory level. Capital deepening or capital intensity (expressed as capital-to-labor ratio) has been one of the conventional indicators of technology adoption.

Determining the Production and Productivity of Organized Pulse Processing: A Regression Analysis

Next, we conduct a rigorous analysis of the organized pulse processing sector using plant-level data. We are interested in assessing the role of state-specific characteristics and learning-by-doing factors in determining mill performance. In particular, we want to investigate the possible role of agglomeration, if any. Do mills that are older tend to be more productive? Does the presence of more mills with larger output in a given neighborhood lead to better outcomes for the average mill in that neighborhood? These questions are important from a policy perspective. If agglomeration matters, then the promotion of new industry might require a critical mass of existing plants in the neighborhood. The answer could also be obtained by assessing the role of state fixed-effects in being associated with mill-level outcomes. Similarly, if experience in the industry is important, then targeting firms or plants with a longer time in operation could be the optimal strategy. Basically, we are interested in estimating the following regression equation involving a repeated cross-section of mills in the organized sector:

$$O_{ijt} = \alpha_j + \beta_t + \vartheta X_{ijt} + \gamma n_{ijt} + \delta Z_{ijt} + \varepsilon_{ijt} \tag{1}$$

In equation 1, O_{ijt} measures a specific outcome for mill i in state j at time t. X_{ijt} is a matrix of characteristics of i in state j at time t. It excludes the two plant- and time-specific variables n_{ijt} and Z_{ijt}, respectively, that are entered as separate variables. n_{ijt} represents the number of years of operation of the mill i in state j at time t. Z_{ijt} is for the agglomeration term; it equals the total output or the number of mills in state j at time t excluding the ith mill. The larger this value is, the greater is the agglomeration factor with its network effect for the ith mill in state j at time t. A significant effect of this variable with the outcome variable would indicate that mills derive benefits from being in a locality that has more mills and/or that are producing larger output. n_{ijt} equals the year data recorded minus the year of inception. It captures the learning-by-doing effects where older plants are more productive (if there is a significant and positive coefficient). α_j and β_t respectively are the state and time fixed effects.

The former captures whether, in relation to a benchmark state, the average outcome of a mill in a specific state is subpar or better. β_t represents the time fixed effect, accounting for generic time factor that affects all mills across all states—for example, when the global food price crisis hit in 2008. These two fixed effects control for state-specific time-invariant unobserved characteristics, such as agroclimatic conditions if they are conducive to pulse cultivation. ε_{ijt} is the classical error term. All standard errors are clustered at the state level. Apart from introducing state and time fixed effects separately, we also include in one specification state × time fixed effects.

In the case of state fixed effects, we take Madhya Pradesh to be the benchmark state, so state-specific effects are measured in relation to Madhya Pradesh. Several unobserved factors, such as governance, are not necessarily time-specific or state-specific but vary across states at different times. A state

× time fixed effect accounts for such factors and minimizes omitted-variable bias. Examples of other variables included in X_{ijt} are the capital-to-labor ratio employed in a specific mill in a state at a particular time, as well as mill- and time-specific use of inputs, mainly fuel and electricity. Because states vary in their degree of urbanization and therefore in the incidence of rural and urban location for the average mill, we also implement a specification with rural fixed effects separately.

Results. Results from estimating equation 1 are presented in Table 5.3. From column 2 on, estimation results correspond to increasing levels of generality. The first specification is the standard linear estimation using ordinary least squares. Subsequently, state and time fixed effects are introduced separately. One specification includes the rural location fixed effects to distinguish the outcomes between rural and urban mills. The coefficients on state fixed effects are presented separately.

A few important points emerge from this estimation:

- In the models the tendency for more experienced firms to be more productive is validated. However, results for agglomeration effects are inconsistent, as one indicator is compatible with the hypothesis of agglomeration effects, while the other indicator points in the opposite direction. With the dependent variable as output per worker, the capital-to-labor ratio is (as expected) strongly associated with higher productivity.
- The significant coefficient on electricity shows the important role it plays in affecting productivity in the pulse processing sector. Enterprise surveys in India often show that energy deficits are a binding constraint. Indeed, energy-deficient states such as Bihar and Uttar Pradesh have a negligible-to-thin spread of the organized pulse processing industry.

The bottom panel in Table 5.3 presents the coefficient of state fixed effects and rural fixed effects, respectively. Results show the following:

- By taking Madhya Pradesh as the benchmark state, the labor productivity in most states with a reasonable presence of pulse processing is subpar to varying degrees. Some frontline states, like Andhra Pradesh and Tamil Nadu, also perform badly in relation to Madhya Pradesh.
- Moreover, there is a clear demarcation in the performance between rural and urban mills; in the organized sector, the average urban mill has significantly higher productivity than the average rural mill.

TABLE 5.3 Determinants of output per worker

Explanatory variable	Linear	State fixed effect	Rural-urban fixed effect	Year fixed effect	State x year fixed effect
Capital-to-labor ratio	0.698***	0.658***	0.710***	0.838***	0.826***
	(0.102)	(0.1000)	(0.100)	(0.0953)	(0.0973)
Petrol	−0.0229	−0.0150	−0.0187	−0.0344	−0.0138
	(0.0238)	(0.0234)	(0.0235)	(0.0221)	(0.0227)
Electricity	0.0173	0.0242**	0.0147	0.0241**	0.0164
	(0.0121)	(0.0121)	(0.0119)	(0.0112)	(0.0129)
Number of years operational	0.000315***	0.000385***	0.000315***	−0.000380	0.000607**
	(5.02e−05)	(5.11e−05)	(4.95e−05)	(0.000752)	(0.000265)
Number of units in the state except this unit	−0.0126***	−0.0156***	−0.0112***	−0.0105***	−0.0169**
	(0.00145)	(0.00350)	(0.00145)	(0.00143)	(0.00838)
Total output generated by the state except this unit	9.27e−05***	6.48e−05***	8.69e−05***	7.15e−05***	−2.39e−05
	(1.01e−05)	(1.21e−05)	(9.96e−06)	(1.03e−05)	(5.11e−05)
Constant	15.05***	15.84***	14.54***	15.57***	16.01***
	(0.0746)	(0.144)	(0.103)	(0.153)	(0.226)
Observations	1,584	1,584	1,584	1,584	1,584
R-squared	0.184	0.244	0.209	0.304	0.409

Source: Data from Annual Survey of Industries.

Note: Standard errors in parentheses. *** $p < 0.01$, ** $p < 0.05$, * $p < 0.1$

State Fixed Effect—Base State: Madhya Pradesh	
Himachal Pradesh	−1.928**
	(0.895)
Punjab	−1.060**
	(0.464)
Uttaranchal	−1.923**
	(0.896)
Haryana	−3.253***
	(0.267)
Rajasthan	−0.694***
	(0.233)

(continued)

TABLE 5.3 (continued)

State Fixed Effect—Base State: Madhya Pradesh	
Uttar Pradesh	−0.499**
	(0.198)
Bihar	−2.786*
	(1.537)
Manipur	−5.137***
Assam	−2.350***
	(0.783)
Orissa	−2.062***
	(0.641)
Chhattisgarh	−1.170***
	(0.235)
Gujarat	−1.101***
	(0.177)
Maharashtra	−0.333*
	(0.181)
Andhra Pradesh	−0.659***
	(0.233)
Karnataka	−0.904***
	(0.234)
Kerala	−1.173
	(1.538)
Tamil Nadu	−1.395***
	(0.207)
Rural-urban fixed effect	
Urban	0.633***
	(0.0907)

Source: Author's calculations.
Note: Standard errors in parentheses.
*** $p < 0.01$, ** $p < 0.05$, * $p < 0.1$

For robustness, Tables 5A.1 and 5A.2 in the chapter appendix present the results of estimation with gross value-added per worker and output per mill as dependent variables. In all of these, our preferred specification is the one containing state × time fixed effects, which best minimizes the possibility of omitted variable bias, but identification of effects is difficult given the limited variation we are left with.

Backward Links: Supply Chains to Pulse Processing

Sufficient availability of pulses for the processing sector is a commonly discussed problem (see, for example, the National Bank for Agriculture and Rural Development [NABARD n.d.]). Banerjee and Palke (2010) examine the supply chain for pulses in general, and Yogan and Manohar (2015) examine the chain for chickpea. A common characteristic of the supply chains involving pulse processing is the large number of intermediaries that lie between the producers and the consumers. These intermediaries include commission agents, wholesalers, processors, and retailers. The marketing channels in pulses are both private and institutional. The institutional arrangement for marketing includes procurement of pulses by providing minimum support prices to the farmers through agencies like India's National Agricultural Cooperative Marketing Federation (NAFED). Under this arrangement, the farmers sell to the procuring agency, and the millers source directly from the procuring agency. In the case of pulse processors, however, the amount of sourcing done through this channel is not significant. The private marketing channel is, in fact, the most common channel, and it exists for pulse processors throughout the country and for every type of pulse.

Supply Chain Channels

Four basic marketing channels for pulses, including processors, are identified by Banerjee and Palke (2010), which may be outlined as follows:

- **Channel 1.** Farmer/producer → village trader → dal miller → wholesaler → retailer → consumer.
- **Channel 2.** Producer → dal miller → retailer → consumer.
- **Channel 3.** Producer → wholesaler → dal miller → retailer → consumer.
- **Channel 4.** Farmer/producer → village trader → commission agent → dal miller → wholesaler → retailer → consumer.

Channel 2, which links producers directly to millers, is comparatively rare. Similarly, millers getting pulses directly from wholesalers is also rare. The more common channels for dal mills are channels 1 and 4. In both these channels the village traders are the initial link in the marketing chain. Traders can buy pulses directly from the farmgate and then supply them to processors, or else they can buy from the *mandis* (government-regulated wholesale markets) through a commission agent. Because there is very limited (if any) procurement by agencies like NAFED, farmers generally sell their pulses in their own villages, in the weekly markets, or in the nearby *mandis*. According to Banerjee and Palke (2010), farmers market about 75 percent of their pulse produce and retain the rest for their own consumption and for seeds for the next year. In the case of chickpea, Yogan and Hansa (2015) look at the supply chain for the processing sector and find similar marketing channels as Banerjee and Palke (2010) do for the other pulses.

Market Power, Price Formation, and Price Transmission in the Supply Chain

In the pulses supply chain involving the processor, with limited or no procurement by the government or directly by the processors (channel 2), most farmers sell to traders. The policy-driven entry barriers in trading mean that traders enjoy certain market power, particularly in relation to the small farmers. For example, to begin business operations to market pulses, any purchaser/dealer/trader needs to take two licenses (India, Ministry of Agriculture 2012):

1. License under the respective state Agricultural Produce Marketing Committee (APMC) Act to deal in agricultural produce (please see below for an explanation about the APMC Act).
2. License to stock pulses under the Essential Commodity Act—Pulses Control Order.

Furthermore, in some states, the traders/commission agents seeking a license are required to have a physical establishment for such business in the APMC market area. These provisions mean that there are comparatively few traders in agricultural commodities in general and pulses in particular, giving traders some degree of monopsony power. Farmers who are located away from the *mandis* (the wholesale markets) usually sell their produce to traders at the farmgate where the farmer's bargaining power is even weaker. The few traders who pick up pulses from the farmgate usually discount the price to include transportation cost to the *mandi*. The problem is compounded because of asymmetric information. Located away from the market, these farmers lack information about prices and are not able to bargain for the best price. The information consists of data and analysis containing inventory, facility, transportation, price, and customers as well (Yogan and Manohar 2015). In the pulses supply chain, including the processors, the buyers and sellers (small farmers) lack information about the external market price given the limited coverage of channel 2. With numerous intermediaries, flow of information and market signals are manipulated that minimize the returns to the farmers and affect the supply to the processors.

Hence, in the *mandis* as well as at the farmgate, one of the results of imperfections is suppression of prices accruing to the farmers. Chapter 3 shows an implicit collusion among traders around the focal point of market support price (MSP). Also, imperfections in the market result in limited transmission of market prices to the farmgate level (see Rahman 2015 for asymmetric price transmission in the case of pulses). Because of low farmgate prices relative to retail prices, we also see small supply responses to rises in pulse prices because farmers receive only a small fraction (less than 50 percent) of what buyers such as processors pay in the market.

Aggregation of the small surpluses through producer companies can possibly help in this context. For example, a recent case study in Tamil Nadu shows that farmers' realization increased from 47 percent to 63 percent of the retail price of white lentils once the growers organized themselves into a producer company (Angles and Karunakaran 2016). A number of farmer producer organizations (FPOs) have been organized for pulse growers across different parts of India, but there is a large variation in their performance. We need more research to understand how to promote successful and viable FPOs that bring more benefits to their members and also help the processors. Successful FPOs will not only help in the marketing of pulses but may also act as effective channels of extension to promote the use of better seeds, lifesaving irrigation, and best practices in pulse production and thereby ensure a consistent supply of good quality raw materials to the mills.

One other change related to marketing that can bring benefits to farmers and millers alike could be to free pulses from APMC taxes. Under the APMC Act, all transactions are regulated to take place in government-licensed wholesale markets (*mandis*). The state governments then impose taxes on all transactions that take place. The buyers have to pay these taxes and they can build it into their bids, which result in price markups. *Mandis* in different states have different taxes. Some states, like Haryana and Uttar Pradesh, have very high taxes (15 percent and 19 percent, respectively). In the major pulse-producing state of Madhya Pradesh, the taxes are as high as 9 percent. These taxes add to increases in consumer prices and reduced farmer prices in pulses, and they should, therefore, be done away with.

Additional Costs through the Supply Chain

The marketing costs in the supply chain leading to the processing sector normally include (1) handling charges at local points, (2) assembling charges, (3) transport and storage costs, and (4) handling charges by wholesalers and retailers. In addition, a market fee is charged either on the basis of weight or on the basis of the value of the produce and is usually collected from the buyers. The seller or the buyer (or sometimes both) pays this commission to the commission agents. On top of handling and marketing fees, across India there are numerous taxes to be paid, such as a toll tax, terminal tax, sales tax, and octroi.[7] All these taxes vary across the markets and from state to state, and the rates are different as well. Usually, the taxes are payable by the seller, but they are built into the prices the processors pay for the raw material. Miscellaneous charges to cover handling, weighing, loading, unloading, and cleaning are payable either by the seller or by the buyer.

Due to this complex set of levies—market fees, commissions, taxes, handling and transport charges, and other miscellaneous charges—the absolute value of the total marketing margin varies widely from market to market, from channel to channel, and from one period to another. For example, in the Azadpur *mandi* in Delhi, officially the commission is 6 percent, but in practice it goes up to 10 percent. In the Vashi market in Navi Mumbai, the officially notified commission is 8 percent, but in practice it runs as high as 15 percent (Banerjee and Palke 2010). Both these markets deal in different types of pulses as well. Based on a field survey, Banerjee and Palke (2010) identified the marketing costs for two types of supply chain involving processors of pigeon pea. They computed the marketing cost and margins for the two most important channels (channel 1 and channel 4) from the producer all the way through the processor (reproduced in Table 5.4).

As Table 5.4 shows, moving from channel 1 to channel 4 adds costs in terms of the *mandi* tax and cess as well as the commission for the agents.[8] These marketing costs are based on sourcing from within the same state where the processor is located. If raw material is brought in from outside the state, other taxes and charges will be applied, compressing the margins further and augmenting the cost of raw materials for the processors.

---

7 Octroi is a tax levied on goods entering a town or city. In India, it is generally imposed by large cities.

8 The government imposes cess (for example, education cess).

TABLE 5.4 Marketing costs–supply chain for pulse processor

	Channel 1			Channel 4	
	Particulars	Rupees per quintal	Percent to the next links purchase price	Particulars	Rupees per quintal
1	Producers' sale price/village traders'purchase price	2,000		Producers' sale price/village traders' purchase price	2,000
2	Cost incurred by producer/farmer			Cost incurred by producer/farmer	
A	Cost of gunny bags	25	1.14	Cost of gunny bags	25
B	Loading	3	0.14	Loading	3
C	Unloading, weighing, and cleaning	8	0.36	Unloading, weighing, and cleaning	8
D	Transportation	39	1.77	Transportation	39
	Total cost (A+B+C+D)	75	3.41	Total cost (A+B+C+D)	75
3	Village traders' margin	125	5.68	Village traders' margin	125
4	Village traders' selling price	2,200	100	Village traders' selling price	2,200
5	Processors' purchase price	2,200	63.77	Processors' purchase price	
				APMC tax and cess	55
				Agents' commission	44
6	Fixed operational cost of dal mill	738.5	21.41	Fixed operational cost of dal mill	738.55
				Total cost	837.55
7	Processors' margin	511.45	14.82	Processors' margin	511.45
8	Processors' selling price	3,450	100	Processors' selling price	3,549

Source: Banerjee and Palke (2010).

Note: APMC tax and cess are imposed by government-regulated wholesale markets.

Constraints in Processing

The major problems for the vast majority of India's present-day pulse processing units are their low product recovery rates and their high milling costs, all stemming from the fact that these processing units are still running on the old, traditional system rather than deploying such modern, sophisticated systems as those used in Australia, Canada, Germany, and Spain (see Banerjee and Palke 2010). In the pigeon pea sample of Banerjee and Palke (2010), almost all the mills were found to be running according to the traditional system, which cause higher milling losses in the form of fragmentation and powder, resulting in a lower recovery of dal than the modern methods.

In addition, the average capacity utilization of these processing units in Banerjee and Palke's sample was just 70 percent, due to the recurring nonavailability of raw pulses and the way processing was operated as a seasonal activity. The units used batch processing, which involves excessive material handling that in turn results in pulse loss. They would prepare a lot of 50 to 60 quintals of pretreated/conditioned pulses at a time for milling, and only after that batch had been converted into dal was the process repeated. Moreover, most of the observed units used sun drying, which reduces their capacity utilization during the rainy season. Moreover, because pulses are aggregated from a large number of players (from either channel), they differ in their quality, variety, and size, but grading and standard-setting for pulses is lacking among the processors. This is because pulses are all mixed together and this can compromise quality. Also, processors are made to sort quality, which imposes costs in terms of time and other resources. In some states, an additional constraint is the lack of an uninterrupted supply of electricity or an uninterrupted supply of water (or both). These are major concerns.

Another constraint is access to financing. The establishment of a dal mill unit involves a large investment in block and working capital. In Banerjee and Palke's sample, the working capital, which constituted around 85 percent to 90 percent of their overall cost of operations, was the most important component. The units obtained loans from informal sources—with interest rates as high as 15 to 20 percent per year—to procure pulses (Banerjee and Palke 2010). Stocking limits represent another constraint. These are limits the government imposes on processors to check speculation and hoarding, and they are often binding. These limits vary enormously from state to state and across time. As of March 2015, for example, Bihar was placing a stocking limit of 1,500 quintals on all pulse mills in the state. In Gujarat, the limit for unmilled pulses was 500 quintals, and 250 quintals for the finished stock of milled pulse. Haryana had a limit of 2,000 quintals. Karnataka put in a regulation requiring that stocks not exceed 30 days' requirement. In Maharashtra the limit for unmilled pulses was equal to one-ninth of a mill's annual production/installed capacity, and for milled pulses it was one-eighteenth of annual production/installed capacity. Other states had varying levels of stock limits. The stocking adversely affects the functioning of pulse processors (Yogan and Manohar 2015).

High price fluctuations in pulses are another constraint, as they discourage processors from stocking larger inventories even within the limit. The government's MSP, which serves only as a benchmark price for pulses (since there is no or limited procurement), has been quite variable over the years. For 2010–2011, 2011–2012, 2012–2013, and 2014–2015, the MSP for chickpea has moved from 2,100 to 2,800, to 3,000, and finally to 3,100 rupees per quintal. With this degree of price fluctuation in raw material prices, staying financially sustainable in the industry becomes a challenge, especially for small-scale millers and manufacturers. There are additional constraints that discourage processors from purchasing outside their own area or state. When millers are located in the same area where the *mandi* is situated, they have the advantage of buying raw materials directly from the wholesale market. But millers located outside a producing state cannot buy products directly; because their inventory is normally limited to a maximum of 10 days' consumption, they do not make bulk purchases partly because there are stocking limits. It is unlikely to be economical to travel to make direct purchases of small quantities, so they must depend instead on agents. As discussed earlier, transacting across state borders also brings in additional costs and charges.

Forward Links between Processor and Consumer

Banerjee and Palke (2010) present data on how the consumer price is arrived at from the processors to the consumers. Going forward from the processors to the consumer, it is either through the wholesalers to the retailers or to retailers directly who then sell to consumers. Banerjee and Palke show that in this chain, processing costs comprise 67 percent to 71 percent of the difference between the farmgate price and the consumer price. In other words, in the chain a sizable portion is contributed by the processing costs. Hence, in the formal or quasi-formal supply chains, significant reduction in consumer prices can be achieved through improved efficiency in pulses processing. As discussed, the pulses supply chain generally has the feature that just when consumer prices are high, the producer prices tend to be low (Yogan and Manohar 2015). Based on Banerjee and Palke (2010) primary data, farmgate prices tend to be on average half of the consumer price. They are a bit higher in channel 4 over channel 2. The lack of direct link between processors and retail is salient here. When the processor sells forward, it includes marketing costs comprising cost of labor, weighing, cleaning, packaging, transportation, and processor's margin. Banerjee and Palke (2010) estimate the wholesale and retail margins to be about 2 percent each in the forward link between processor and consumer. Bantilan et al. (2014) estimate large effects of improved technical efficiency in chickpea processing on consumer prices. Hence supply chain efficiencies through improved processing and better production technology for growers can likely bring down consumer prices of pulses.

Policies for Improving Pulse Processing

Policies need to be designed to address several issues confronting the pulse processing sector.

Addressing the supply problem. The uninterrupted supply of raw material is a prerequisite for running pulse processing units efficiently. Yogan and Manohar (2015) and Banerjee and Palke (2010) both show that this has been a binding factor for pulse processing. Apart from resorting to imports, processing units cannot maintain continuous operations all year round because the available domestic production has generally been inadequate. Efforts should, therefore, be made to ensure a supply of raw material throughout the year. Supply chains need to be developed adequately. In the case of Tamil Nadu's processing mills, for example, chickpeas must be supplied from Andhra Pradesh, Karnataka, Madhya Pradesh, and Maharashtra. In addition, imported raw material is purchased from both private parties and from government institutions. Relying on multiple sources has been the only solution for several processing units to meet their requirements for raw material. In this context, systems that would encourage direct purchase by the processors from farmers could be helpful—that is, after diluting APMC restrictions. Direct marketing would enable farmers as well as processors to economize on transportation costs and also improve price realization. Because pulses are sourced from a large number of small farmers, there is a need to establish a system of grading and sorting to standardize the inputs.

Addressing farmers' marketing challenges. Along with the buyers, farmers face such problems as delayed payments and lack of bargaining power as compared with licensed traders as well as a lack of adequate infrastructure in the *mandis* from where the processors source their pulses. Even in states that have diluted or done away with the APMC Act, new markets have not replaced them, forcing most farmers to sell their produce to traders who, as aggregators, supply to different buyers, including processors.

Addressing inefficiencies in mill performance. There are avenues for exploiting new technologies in pulse processing, some of which have been developed in India itself, as identified in studies like those by NABARD and Tamil Nadu Agricultural University (TNAU). For example, institutes like

Central Food Technological Research Institute (CFTRI) in Mysore have developed a conditioning technique to loosen the husk without resorting to sun drying and oil and water application. This step has been mechanized with the introduction of conditioning units. Use of this conditioning technique, as developed by CFTRI-Mysore, could be one option for improving mill performance.

The present rate of losses can be greatly minimized through the use of improved dal mills. Several research institutes in India have developed improved dal mill technologies that are highly versatile and energy efficient.[9] The improved dal mills have a dehusking efficiency of about 95 percent, and their yield of split pulses is about 80 percent to 85 percent (Banerjee and Palke 2010). Greater use of these technologies could be helpful for the growth and sustainability of pulse processing.

Addressing marketing needs of the processors. Finally, in marketing, unlike other branded products such as basmati rice and edible oils, apart from *besan* no branded product of any pulse is currently popular (Banerjee and Palke 2010). Processed pulses have seen a significant spread of branded products in recent times, with processing companies like Halidram and Bikano, but in dal the branding and product differentiation has so far been limited. How to improve branding and product differentiation in dal is an important problem to address.

Consolidation of Firms and Organizing of Farmers

The problems that the processing sector has faced due to both irregular supply of raw materials and inadequate quality (leading to excess capacity) can be mitigated by better coordination with the farmers (Banerjee and Palke 2010). The analysis in this chapter shows that there is a gradual process of consolidation happening on the firm side, with the processing sector undergoing scaling up and concentration of firms. The growth in sales of processed food has also led to an increase in processing firms' use of pulses. The path to firm consolidation in processing is likely to follow three stages (Bora, Gulati, and Roy 2006). In the initial stage, small processors dominate, but with increasing income and urbanization, changes in food habits, and the entry of foreign direct investment (FDI), firms can scale up their operations. New firms

9 These include PDKV in Akola, CFTRI in Mysore, Gobind Ballav Pant University of Agriculture and Technology (GBPUAT) in Pantnagar, CIAE in Bhopal, Indian Institute of Pulse Research (IIPR) in Kanpur, Tamil Nadu Agricultural University (TNAU), and Coimbatore and Indian Agriculture Research Institute (IARI) in New Delhi.

come in with product differentiation, and a constant process of churning is observed. However, by the third and final stage only the most efficient firms—those that can adapt to the demands of the market—survive, leading to a high level of concentration with a few scaled-up firms dominating the market. In India, the empirical evidence shows that there has been a definite scaling up in the food-processing sector (Bhavani, Gulati, and Roy 2006). As this chapter has shown, the levels of output and capital per firm have gone up substantially in pulse processing since 2001. It seems that pulse processing is currently in stage 2. With the right policy support, a transition to stage 3 can happen, and if it is managed well, it can be an opportunity to make pulse processing dynamic.

In dealing with the farmers, who are mostly smallholders, creating scale is important. Pulse growers are increasingly organizing themselves into farmer producer organizations (FPOs), especially in the states that are leading producers, such as Madhya Pradesh and Maharashtra. Indeed, the government's integrated program for the development of 60,000 pulse villages, which is managed by the Ministry of Agriculture through its Small Farmers Agribusiness Consortium (SFAC), explicitly lists the promotion of FPOs in the pulse sector as a policy to improve outcomes. The SFAC argues that to work out economies of scale and link pulse farmers to markets, it is necessary to organize pulse farmers into groups, both for getting access to quality inputs and for creating market links, such as those with processors. The government announced 2014 as "the year of FPOs" and several FPOs in pulses were initiated (India, Ministry of Agriculture 2013).

To date, though, there have been only a few cases of contract farming by the pulse processing sector. Examples include the Odisha Rural Development and Marketing Society (ORMAS) for pigeon pea. Under this arrangement, which is managed by self-help groups, the processor has an assured market by supplying pulses for food programs like the Mid-Day Meal Scheme. This initiative has resulted in ensuring a good income to the members of the self-help groups. In 2010, ORMAS procured about 6,000 quintals of local pigeon pea through these groups (Sharma 2010). A private-sector example of contract farming in pulses processing is that of the Tata Chemicals Company (through its subsidiary, Rallis India), which markets differentiated pulses after processing by keeping them unpolished. The product is sold with its brand of pulses, called i-Shakti. Tata Chemicals purchases pulses largely from the Wardha Cotton and Soya Producer Company, which has a mill in Karnataka and recently opened one in Maharashtra.

There is also the case of an NGO that began contracting for pulse processing in 2009. The NGO, known as Seva Mandir, is based in Udaipur in Rajasthan and its aim is to make farmers independent of the intermediaries for selling their produce. The arrangement comprises a cluster of seven villages where the production of pigeon pea is significant. The processing mill is run by a farmer producers' group consisting of local pulse farmers, while Seva Mandir has played a major role in marketing the produce. Currently, the size of the farmer group is small, comprising 86 farmers, but access to the mill is not limited to members. Other pulses procured for processing are green gram and black matpe, and the NGO uses prices in neighboring wholesale markets as reference prices. Like the Tata i-Shakti example, the Seva Mandir processing arrangement is one based on introducing product differentiation and branding. Both arrangements maintain that the pulses they market must be free of chemical fertilizers and pesticides, and both marketed them as premium organic produce.

It is notable that in the Gulbarga district of Karnataka, the "pigeon pea bowl" of India, where there are nearly 300 mills and aggregate production can be high, there are very few instances of contract farming. Instead, most processors buy from government-regulated wholesale markets. Karnataka state government has taken a step forward, nevertheless, by introducing *e-tendering*, and it has developed a common market across the state that can help the processors meet the scale requirement and get good-quality raw material (see Chengappa et al. 2012 and Athawale 2014 for details). As pulse processing has come up in the adjoining states of Andhra Pradesh and Maharashtra (with a large number of organized sector mills), the Gulbarga processing sector has shrunk a bit.

Conclusion

This chapter presented an analysis of pulse processing using secondary-data sources for 2002–2012 for both the organized and the unorganized sectors. Several key messages emerge from the analysis. First, although the unorganized sector dominates in pulse processing in terms of the number of mills and employment, the organized sector dominates in terms of sales value. The analysis shows that a shift is taking place, away from the unorganized sector and moving toward the dominance of the organized sector. Significant restructuring is clearly taking place, with the proportion of large mills increasing, bringing a corresponding increase in installed capacity. This shift, however, is more pronounced in some states than in others, likely because of learning effects, in addition to the individual state's supporting infrastructure. State-specific factors seem to be quite important in determining the sector's outcomes. The pulse processing sector does not seem to be a big generator of employment, either in the organized or the unorganized sector. The major employment generation that does occur is located at the back end—that is, in pulse production, which gets a strong boost from the existence of processing links. For these links, the quality of inputs and their consistent supply remain as concerns. The level of technology used in pulse processing needs to be raised, particularly in the unorganized sector. When compared with other countries, India's pulse processing sector has a low capital-to-labor ratio and only moderate capital deepening over time. In many Southeast Asian countries, small-scale rural food-processing industries have functioned as growth engines (Sharma, Panthania, and Vashist 2003). Right now, the demand for processed pulses in India remains primarily for items that require only primary processing, implying that value-addition in pulses through secondary processing has been limited.

Indian consumers are highly price-sensitive, so a reduction in the cost of processed pulses through more efficient processing is needed to raise demand and consumption. Currently, there is low capacity utilization, and that in turn leads to higher processing costs. Although we do not have data on the exact capacity utilization in pulse processing, based on the standard in the Indian food-processing sector as a whole, we can estimate that it is less than 70 percent. Part of the reason for the lack of capacity utilization is structural, with the seasonal nature of the activity itself leaving mills idle except when imports or stored inputs are channeled to run the mills uninterrupted. There might be limits to scaling up the pulse processing sector. Before it can become an engine of growth, it must undergo a structural shift toward larger mills with higher productivity and with institutional arrangements that can ensure regular supply of good quality pulses. According to Banerjee and Palke (2010), to add value to the pulses, the processing units need to be equipped so as to meet consumer demand. They recommend improving the supply of good-quality raw materials, adopting modern conditioning techniques to loosen the husk without resorting to sun drying, and extending support for storage facility construction and other infrastructure.

The few recently introduced models of Tata i-Shakti, on a larger scale, and of Seva Mandir, on a smaller scale, may offer insight into ways the processing and marketing of pulses as health food may help realize the potential of pulse processing. With the changes under way in the Indian food system, what is needed is a new way of doing business, a new approach based on innovative institutions that can cut transaction and marketing costs for both firms and farms. There is a need to scale up success stories like that of the Indian dairy cooperative movement, where the processing sector played a pivotal role. As suggested by the chief economic adviser to the government of India, there might be a need to apply the Amul model for pulses.[10]

The solutions also require that government play a complementary role in the building of appropriate infrastructure and institutional support. Institutional bottlenecks require significant policy changes to induce efficiency. For example, the taxation rate on processed foods in India is one of the world's highest. Dev and Rao (2004) state that the net tax level is 21–23 percent on food items in India, while the comparative tax burden is 10 percent in the Philippines, Indonesia, and Malaysia; 14–15 percent in the Netherlands and the UK; and 17 percent in China and Ireland. It is also necessary that a uniform value-added tax is imposed on all states to facilitate growth. The current move toward a Generalized System of Taxes (GST) is a big policy step that is likely to boost the pulse processing sector. Similarly, the recent announcement regarding a National Agricultural Market could help processors source materials from a larger area. Several states have started amending the APMC Act, which is crucial for market reform. Toward that end, India's Ministry of Agriculture had formulated a model law on agricultural marketing in consultation with state governments, which enables the establishment of private markets/yards, direct purchase centers, consumers/farmers markets for direct sale, and promotion of public-private partnerships (PPPs) in the management and development of agricultural markets. The regulation and promotion of contract farming arrangements are part of this legislation.

References

ASI (Annual Survey of Industries). 2001–2015. India, Ministry of Statistics and Programme Implementation. Central Statistics Office. Industrial Statistics Wing. www.csoisw.gov.in/cms/En/1023-annual-survey-of-industries.aspx.

Athawale, G. 2014. "APMC and E-trading for Financial Inclusiveness in Karnataka." *IBMRD Journal of Management and Research* 3 (2): 84–98.

10 The Amul model involves firm-farm links where the state-run dairy cooperatives link with dairy processing. This system has been largely credited with making India the world's largest producer of milk. See "Amul Model" 2015.

"'Amul Model' Needed for Pulses: Arvind Subramanian." 2015. Accessed February 2016. www.thehindu.com/news/national/other-states/amul-model-needed-for-pulses-arvind-subramanian/article7904394.ece.

Angles, S., and K. R. Karunakaran. 2016. "Value-Chain Development for Blackgram in Tamil Nadu through Group Marketing." Paper presented at the Conference on Pulses for Sustainable Agriculture and Human Health, New Delhi, May 31–June 1.

Banerjee, G., and L. M. Palke. 2010. "An Overview of Pulses." In *Economics of Pulses Production and Processing in India,* 1–18. Department of Economic Analysis and Research, National Bank for Agriculture and Rural Development (NABARD), Occasional Paper 51.

Bantilan, M. C. S., D. Kumara Charyulu, P. Gaur, D. Moses Shyam, and J. S. Davis. 2014. "Short Duration Chickpea Technology: Enabling Legumes Revolution in Andhra Pradesh, India." Research Report 23, Patancheru, India, ICRISAT. www.icrisat.org/ what-we-do/mip/SPIA .pdf. Accessed September 2016.

Bhavani, T. A., A. Gulati, and D. Roy. 2006. "Structure of the Indian Food Processing Industry: Have Reforms Made a Difference?" In *Plate to Plough: Agricultural Diversification and Its Implications for the Smallholders in India.* Submitted to Ford Foundation. Washington, DC: International Food Policy Research Institute (IFPRI).

Bora S., A. Gulati, and D. Roy. 2006. "Fragmenting Bottom and Consolidating Top: India's Changing Food System and Implications for Small Holders." Mimeo. International Food Policy Research Institute (IFPRI), Washington, DC.

Chavan, J. K., D. M. Shere, H. K. Jawale, and D. K. Salunkhe. 1983. "Effects of Soak Treatment to Legume Seeds on the Cooking Quality of Resultant Dal." *Indian Journal of Nutrition and Dietetics* 20: 249–254.

Chengappa, P. G., M. Arun, C. G. Yadava, and H. M. Prasanna Kumar. 2012. "IT Application in Agricultural Marketing Service Delivery—Electronic Tender System in Regulated Markets." *Agricultural Economics Research Review* 25: 359–372.

Dev, M. S., and N. C. Rao. 2004. "Food Processing in Andhra Pradesh Opportunities and Challenges." Working Paper 57. Centre for Economic and Social Studies, Begumpet, Hyderabad.

Geervani, P., and F. Theophilus. 1980. "Effect of Home Processing on the Protein Quality of Selected Legumes." *Journal of Food Science* 42: 1269–1275.

Gulati, A., P. K. Joshi, and M. Landes. 2008: "Contract Farming in India: An Introduction." Accessed June 2016. www.ncap.res.in/contract_%20farming/Resources/1.Introduction.pdf.

ICRISAT (International Crops Research Institute for the Semi-Arid Tropics). 1991. *Uses of Tropical Grain Legumes.* Proceedings of a Consultants Meeting, March 27–30. Accessed September 2016. http://oar.icrisat.org/964/1/RA_000177.pdf.

India, DSIR (Department of Scientific and Industrial Research). 2007. "Food Processing—Indian Regulatory and Policy Environment." Accessed June 2016. www.dsir.nic.in/reports/isr1/Food%20Processing/6_5.pdf.

———. 2012. "Expert Group on Pulses (Alagh Committee Report), 2012." Ministry of Agriculture, Department of Agriculture and Cooperation (DAC), New Delhi.

———. 2013. *Policy and Process Guidelines for Farmer Producer Organizations*. Ministry of Agriculture, Department of Agriculture and Cooperation (DAC), New Delhi. Accessed June 2016. http://nhm.nic.in/Archive/FPO-Policy&Process-GuidelinesDAC2013.pdf.

Jambunathan, R., S. D. Hall, P. Sudhir, V. Rajan, and V. Sadhana, 1991. *Uses of Tropical Grain Legumes*. Proceedings of a Consultants Meeting, March 27–30, 1989. International Crops Research Institute for the Semi-Arid Tropics.

NAAS (National Academy of Agricultural Sciences). 2006. "Employment Opportunities in Farm and Non-Farm Sectors Through Technological Interventions with Emphasis on Primary Value Addition." Policy paper 37. Accessed July 2016. http://naasindia.org/Policy%20Papers/policy%2037.pdf.

NABARD (National Bank for Agriculture and Rural Development). N.d. *Knowledge Banks. Dal Mills*. Accessed July 2016. https://www.nabard.org/english/dal.aspx.

Parpia, H. A. B. 1973. "Utilization Problems in Food Legumes." In *Nutritional Improvement of Food Legumes by Breeding*, 281–295. Proceedings of a symposium, July 3–5, 1972, Rome, Italy. New York, NY, US: Protein Advisory Group.

Rahman, A. 2015. "Is Price Transmission in the Indian Pulses Market Asymmetric?" *Journal of Quantitative Economics* 13 (1): 129–146.

Rama Rao, G. 1974. "Effect of Heat on the Proteins of Groundnut and Bengal Gram (*Cicer arietinum*)." *Indian Journal of Nutrition and Dietetics* 11: 268–273.

Sharma, K. D., M. S. Pathania, and G. D. Vashist. 2003. "Role of Rural Women in Small Scale Agro-Processing Sector: An Economic Analysis of Samridhi Mahila Processing Co-operative Society in Himachal Pradesh." *Indian Journal of Agricultural Economics* 58 (3): 578–588.

Sharma, P. 2010. *A Study on Agricultural Marketing System in Odisha*. Jaipur, Rajasthan, India: National Institute of Agricultural Marketing (NIAM).

TNAU (Tamil Nadu Agriculture University) Portal. 2016. "Post Harvest Technology: Agriculture. Pulses." Accessed September 2016. www.agritech.tnau.ac.in/postharvest/pht_pulses.html.

Vijaya Kumar, M., ed. 2006. "Employment Opportunities in Farm and Non-Farm Sectors through Technological Interventions with Emphasis on Primary Value Addition." Policy paper 37. National Academy of Agricultural Sciences (NAAS), New Delhi, October.

Yogan, V., and M. L. Hansa. 2015. "Supply Chain Management of Pulses (Chickpeas) Industry in India." *International Journal of Economic Research* 12 (2): 319–329.

Yogan, V., and H. L. Manohar. 2015. "Supply Chain Management of Pulses (Chickpeas in India)." *International Journal of Economics Research* 12 (2): 319–329.

Appendix

TABLE 5A.1 Determinants of gross value-added per worker

Explanatory variable	Linear	State fixed effect	Rural urban fixed effect	Year fixed effect	State cross year fixed effect
Capital-to-labor ratio	0.956***	0.936***	0.964***	0.924***	0.899***
	(0.0716)	(0.0706)	(0.0714)	(0.0718)	(0.0756)
Petrol	−0.0180	−0.0171	−0.0164	−0.0156	−0.00369
	(0.0157)	(0.0154)	(0.0157)	(0.0157)	(0.0163)
Electricity	0.0269***	0.0280***	0.0259***	0.0254***	0.0224**
	(0.00795)	(0.00805)	(0.00793)	(0.00793)	(0.00947)
Number of years since operational	0.000261***	0.000315***	0.000260***	−0.000424	0.000346*
	(3.33e−05)	(3.41e−05)	(3.32e−05)	(0.000511)	(0.000188)
Number of units in the state except this unit	−0.00934***	−0.0108***	−0.00880***	−0.00800***	−0.0172***
	(0.00101)	(0.00233)	(0.00102)	(0.00107)	(0.00593)
Total output produced by state except this unit	6.13e−05***	3.53e−05***	5.90e−05***	5.22e−05***	3.08e−05
	(6.74e−06)	(8.37e−06)	(6.76e−06)	(7.27e−06)	(3.70e−05)
Constant	12.14***	12.60***	11.97***	12.29***	12.65***
	(0.0508)	(0.0966)	(0.0723)	(0.107)	(0.166)
Observations	1,326	1,326	1,326	1,326	1,326
R-squared	0.346	0.392	0.351	0.358	0.438

State Fixed Effect: Base State—Madhya Pradesh		
Jammu and Kashmir	−2.595***	
	(0.969)	
Punjab	−0.741**	
	(0.320)	
Uttaranchal	−2.424***	
	(0.566)	
Haryana	−0.941*	
	(0.485)	
Rajasthan	−0.526***	
	(0.161)	
Uttar Pradesh	−0.510***	
	(0.133)	
Manipur	−2.201**	
	(0.969)	
Assam	−1.543***	
	(0.495)	
Jharkhand	1.457**	
	(0.566)	
Orissa	−1.439***	
	(0.492)	
Chhattisgarh	−0.760***	
	(0.166)	
Gujarat	−0.514***	
	(0.120)	
Andhra Pradesh	−0.422***	
	(0.154)	
Karnataka	−0.448***	
	(0.160)	
Tamil Nadu	−0.640***	
	(0.136)	
Urban		0.207***
		(0.0638)

Source: Authors' estimations.

TABLE 5A.2 Determinants of output per mill

Explanatory variable	Linear	State fixed effect	Rural–urban fixed effect	Year fixed effect	State cross-year fixed effect
Capital-to-labor ratio	0.563***	0.531***	0.571***	0.710***	0.725***
	(0.111)	(0.109)	(0.110)	(0.104)	(0.106)
Petrol	0.0242	0.0350	0.0272	0.0114	0.0436*
	(0.0259)	(0.0254)	(0.0258)	(0.0242)	(0.0246)
Electricity	0.0286**	0.0427***	0.0267**	0.0359***	0.0265*
	(0.0132)	(0.0131)	(0.0131)	(0.0123)	(0.0140)
Number of years operational	0.000257***	0.000314***	0.000256***	−0.000317	0.000836***
	(5.46e−05)	(5.54e−05)	(5.42e−05)	(0.000823)	(0.000288)
Number of units in the state except this unit	−0.0153***	−0.0154***	−0.0143***	−0.0135***	−0.0235**
	(0.00158)	(0.00380)	(0.00159)	(0.00157)	(0.00910)
Total output produced by the state except this unit	0.000106***	7.89e−05***	0.000102***	8.47e−05***	−6.81e−05
	(1.09e−05)	(1.32e−05)	(1.09e−05)	(1.13e−05)	(5.55e−05)
Constant	17.25***	17.76***	16.89***	17.78***	18.25***
	(0.0811)	(0.157)	(0.113)	(0.167)	(0.246)
Observations	1,584	1,584	1,584	1,584	1,584
R-squared	0.197	0.258	0.207	0.305	0.419

State Fixed Effect: Base State—Madhya Pradesh		
Himachal Pradesh	−2.291**	
	(0.972)	
Uttaranchal	−1.984**	
	(0.973)	
Haryana	−4.816***	
	(0.711)	
Delhi	−0.200	
	(0.290)	
Rajasthan	−0.442*	
	(0.253)	
Bihar	−3.607**	
	(1.669)	
Manipur	−5.959***	
	(1.669)	
Assam	−2.484***	
Orissa	−2.335***	
	(0.696)	
Chhattisgarh	−0.920***	
	(0.256)	
Gujarat	−0.583***	
	(0.193)	
Andhra Pradesh	−0.676***	
	(0.253)	
Karnataka	−0.498*	
	(0.254)	
Kerala	−0.000906	
	(1.669)	
Tamil Nadu	−1.044***	
	(0.225)	
Rural-urban fixed effect		
_IRU_Code_2		0.454*** (0.0994)

Source: Authors' calculations.

Note: Standard errors in parentheses. *** $p < 0.01$, ** $p < 0.05$, * $p < 0.1$

Chapter 6

DYNAMICS OF PULSES TRADE IN INDIA

Raj Chandra, P. K. Joshi, Akanksha Negi, and Devesh Roy

During the past decade, because of the persistent deficit in India's pulse sector, the sharp rise in pulse prices has coincided with significant changes in the nature and extent of the pulse trade. Surprisingly, no systematic study has been done on the evolution of this trade, despite the fact that the role of trade in the price formation of pulses is becoming increasingly important. This chapter attempts to fill this research gap. Since exports of pulses from India are comparatively small, we focus on imports.

Background

As discussed in Chapters 2 and 3, India grows and consumes several types of pulses primarily because of heterogeneity in preference across regions. The production and consumption data for the period 2000–2012 show an estimated annual shortage of 2 million to 3 million tons of pulses, which has ushered in significant imports. How the trade in pulses evolves has significant implications for their production, consumption, prices, and generally on the entire supply chain, including processing and retailing. Imports fill the gap between production and consumption and can help cool inflationary pressure (Gokarn 2011). Although pulses do not have a comparatively high weight (vis-à-vis commodities like cereals and some animal products) in the wholesale price index used to measure inflation in India, pulses have experienced high prices for a long time. This has been particularly true since 2005, so that pulses figure in the list of commodities that have driven the rise in the relative price of food.

Trade as a tool for stabilizing prices has been employed across a spectrum of food products facing price pressures, and pulses in India are no exception (see Minot 2011 for examples). In many cases where there is a threat of significant price rise, particularly in cereals, the trade policy response has been quick and large. For example, when there was pressure on wheat prices in 2006, India quickly arranged imports of 6 million metric tons of wheat from Australia

(Murugkar 2006).[1] In addition, stop-gap imports of milk, sugar, and certain vegetables (such as onion) represent a common policy stance in response to rising prices. The liberal trade policies adopted for pulses since 2000 were probably adopted under the same principle. In response to such policies and the expanding set of trading partners who see India's pulse market as a significant opportunity, pulse imports have been steadily growing, rising by as much as 36 percent in the previous decade beginning in 2001. Although the evolution of pulse imports has coincided with a parallel increase in their domestic prices, to the best of our knowledge, no rigorous study of pulse imports and the extent to which they cool the domestic pulse markets has been undertaken.

Study Objectives and Data Sources

To study these dynamics and test the hypothesis of a cooling effect, we use a unique dataset from the government of India's Customs Department that provides disaggregated trade flow data at high frequency.[2] High frequency is extremely important since annual data—which is how most trade statistics are reported—cannot be used to study the dynamic price behavior of pulses and its relationship with imports. In addition to revealing details on landing sites, types of pulses, and unit values, the customs data allow the dating of imports. Only by dating imports can we establish any concordance between a high frequency of imports and a similar frequency in price indexes.

To study the impact of imports on domestic prices, we need to map pulse imports onto their counterparts in the domestic price data at identical or at least similar frequencies. Employing the wholesale price data and the port data from the Customs Department, we assess the cooling effect of pigeon pea imports, one of the most important pulses in India in terms of production and consumption. This pulse has also been subjected to persistent price spikes.[3] Our study has the following objectives: (1) to describe the evolution of India's pulses imports disaggregated by commodities; (2) to identify the time-series behavior of different types of pulse imports and possible substitution between

1 Hereafter, measurements in tons are metric tons. India has been self-sufficient in wheat for a long time (since 1980). However, India imports wheat during years of unfavorable weather to meet the growing demand and also to add to stocks, as was the case in 2006. Between 2012 and 2014, India reached wheat production volumes of 95 million tons and was second only to China in production.

2 Cooling effect refers to pulses imports possibly reducing short-term upward price pressure in India's domestic market.

3 In recent years the retail price of pigeon pea surged to 120 rupees per kilogram, whereas other pulses remained at around 70 rupees per kilogram for more than six months (Reddy 2009a).

the imports and domestic sources, if any; (3) to analyze the changes in imports along the extensive (expansion of the set of importers) and intensive (expansion of imports per importer) margins of the pulse trade; and (4) to answer these questions:

- How have imports of pulses affected domestic prices?
- What is the nature of the effect on prices—that is, do imports bring prices down or just slow their rate of increase?
- What is the time span for imports to have a bearing on the price of pulses?

As mentioned, the study employed highly disaggregated customs data that allow a finer classification of product and source (up to the company level). The rich dataset, available at the 8-digit Harmonized System level (HS), helped in documenting stylized facts on the evolution of the pulse trade with comparatively high accuracy and significant details. The HS system developed by the World Customs Organization (WCO) is an international standard system for classifying traded products. Under the classification, pulses come under chapter 07 (subheading 0713 at the four-digit level), which includes such items as "dried leguminous vegetables, shelled, split or skinned." Table 6.1 lists the broad categories of pulses under the HS classification with their codes.

The time period of our study is 2001 to 2012. The customs data contain information on both import and export values (in current US dollars and in Indian rupees) and volumes of pulses traded. However, the data on volumes were not originally standardized in terms of units; rather, several types of units were used such as containers, bags, kilograms, and boxes, making it difficult to bring them under a common denomination. Hence, we standardized the quantities based on unit values from non-missing data specified in metric units. Using unit values and identifying commonality in value-unit relationships, the units were standardized to kilograms for all quantity flows. Several validation checks were done to ensure that at least in orders of magnitude there was comparability with other data, wherever available. The pulses imports were grouped under seven main categories: pea, pigeon pea, chickpea, green gram, black matpe, lentils, and beans.[4] The category "pea" included yellow pea, split yellow pea, green pea, and dry pea. The pigeon pea group

4 01731000 represents peas, which includes both yellow and green pea. However, customs data also provide product description along with an HS code. So, even if both yellow pea and green pea came under the same HS code, with the help of product description, we could separate out the data for yellow pea and green pea. The same was the case with black matpe and green gram.

TABLE 6.1 Classification of pulses under the Harmonized System

HS Code	Description
07131000	Peas
07132000	Chickpea
07133100	Beans of the species *Vigna mungo* (L) Hepper or *Vigna radiate* (L) Wilczek
07133200	Small red (Adzuki) beans
07133300	Kidney beans, including white pea beans
07133400	Bambara beans
07133500	Cowpeas
07133910	Guar seed
07133990	Others
07133400	Lentil
07135000	Broad beans and horse beans
07136000	Pigeon pea
07139010	Split
07139090	Other

Source: Data from UN Comtrade database.

included whole as well as split pigeon pea. Beans as a group constituted a small portion of the total pulses trade. All types of beans were therefore combined into one category, which included kidney beans, black-eyed beans, cowpea, lablab beans, and green beans. Cumulatively, different beans accounted for 5 percent to 6 percent of the total pulses imports.

The customs data are recorded at a daily frequency. For this study, the daily imports were converted into weekly imports by taking the average of daily imports over seven days. This was done to include days when there was no port arrival and also to match the import data with price data recorded at a weekly frequency. The Wholesale Price Index (WPI) for pulses was used for weekly domestic prices. The weekly import values (in millions of [constant] US dollars) and the WPI of pulses were used for the period 2002–2012.

Import Policies Affecting Pulses

During the 1970s and 1980s, India followed a protectionist trade policy with respect to agricultural commodities. Imports were restricted with the aim of safeguarding and promoting the interests of domestic agriculture. The Indian government tried to achieve this by imposing quantitative restrictions, quotas, tariffs, and a variety of other equally prohibitive trade mechanisms (Agbola

2003). All potentially importable items in India were categorized under three heads: the prohibited list, the special list, and the free list. Earlier, pulses had been on the special list, meaning their import was permitted subject to licensing. In the 1990s, India undertook structural reforms and adopted a more liberal outlook on international trade, leading to significant reductions in tariff and nontariff barriers. This liberalization, however, was mostly for nonagricultural products, implying that domestic terms of trade improved for agriculture.

Before liberalization, although the import of most agricultural commodities was subject to licensing and quantitative restrictions, India's import policy in pulses was comparatively liberal. In 1979 the import of pulses was placed under Open General License (OGL), which made it possible for any public- or private-sector entity to import without approval or any restriction. The import duties on pulses declined steadily during the 1980s and 1990s (Price, Landes, and Govindan 2003). From 1989 to 1994 the import duty on pulses was only 10 percent. This was further reduced to 5 percent in 1995, and it was eliminated entirely in 2000. In 2001, a duty of 5 percent was again placed on pulses, and in 2002–2003 that was increased further to 10 percent (Sathe and Agarwal 2004). From 2007 to 2012 imports of pulses were made duty free, and in 2013 the duty on imports was reduced to zero (India, Ministry of Agriculture 2013).

However, from the perspective of importers and traders, it was understood that having a 0 percent import duty did not ensure that the pulse market would remain liberal in the future. Agricultural trade policy in India has always been variable and thus uncertain. Consequently, even when there is near free trade in pulses, agents might not expect that situation to persist in the future. It could be that the only reason imports are enjoying freer access to the market in India is because imports have not as yet reached levels that threaten the domestic sector. It is understood that if such a (threatening) situation emerged, the government might respond by tightening import policy.

Changing Patterns in Pulse Prices, Consumption, and Imports

The import prices and the domestic wholesale prices of pulses tend to move in parallel, as both Figure 6.1 and Table 6.2 illustrate. (Figure 6.1 illustrates the pairing for pigeon pea, and Table 6.2 pairs the average annual domestic prices with the import unit values for different pulses over the study period.) During the 2008 food price crisis, for example, both the domestic price and the import price of pigeon pea—which is explored further below to examine

FIGURE 6.1 Weekly imports and wholesale prices (WPI) of pigeon pea, 2002–2012

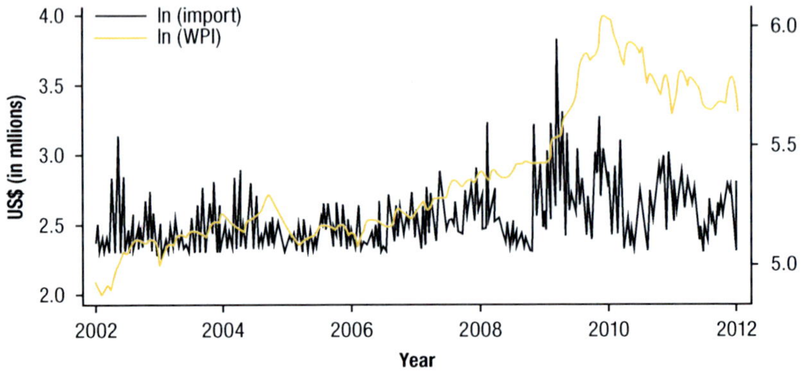

Source: India Customs Data (2002–2012).
Note: Imports are in 2005 constant US$ (in millions), shifted up by 10 units.

its cooling effect—rose. Moreover, it is evident that for pulses that are traded more, such as lentils, the domestic and import prices have an even closer parity. In contrast, for green gram the gap between the domestic and import prices is starker. In most years the import prices of yellow pea are strikingly lower than those for all other pulses; note that yellow pea has no domestic production in India, and it is not imported as a separate product but, rather, arrives mixed with other pulses.

Even though household consumption of pulses has been declining in India, the country's rising population and incomes have meant that the overall demand for pulses has grown over time. NSS data show that from 1988 to 2009, the per capita consumption of all pulses, except yellow pea, declined. For yellow pea, per capita consumption over that period increased substantially—by 73 percent. Peas have had the largest share in pulse imports, and yellow peas in particular, which are not grown in India and come largely from Canada, have the highest import penetration among all pulses. Nevertheless, although the per capita consumption of yellow pea rose significantly, its share in total pulse consumption has not been high. Rising from a low base, its share out of all pulse consumption was just 5 percent.

Figure 6.2 presents the production, consumption, and imports of pulses in India over our study period (2000–2012). There has been a secular trend (positive) in the import penetration of pulses. Imports constituted less than 1 percent of the total pulse consumption in 2000 and then shot up to 19 percent to 20 percent of consumption within a decade.

TABLE 6.2 Average annual domestic prices and import unit value for different pulses, 2002–2012 (US$)

	Pigeon pea price		Chickpea price		Black matpe price		Lentils price		Green gram price		Yellow pea price
Year	Import	Domestic	Import	Domestic	Import	Domestic	Import	Domestic	Import	Domestic	Import
2002	328	525	352	339	300	514	291	307	382	402	261
2003	301	536	385	333	234	434	408	365	339	374	246
2004	484	620	424	340	310	467	407	457	430	354	257
2005	533	630	477	388	374	571	386	449	474	479	311
2006	477	509	565	497	666	883	475	412	707	618	384
2007	854	609	601	540	1,046	919	506	621	1,084	597	470
2008	1,102	684	649	526	613	777	1,012	840	684	576	659
2009	1,092	1,169	528	467	944	892	1,016	839	918	812	475
2010	1,280	1,339	640	491	1,189	1,178	950	740	1,494	1,005	524
2011	888	1,130	807	609	923	1,124	689	594	1,138	820	464
2012	766	937	852	778	668	937	623	614	890	788	436

Source: Domestic price, Center for Monitoring of Indian Economy (CMIE).

Note: Import price/Unit Value, calculated from Customs Data.

FIGURE 6.2 Production, consumption, import, and import penetration of pulses, 2000–2012

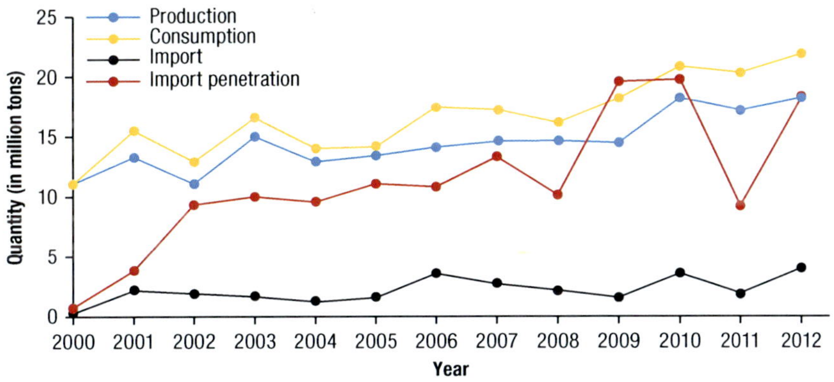

Source: India, Ministry of Agriculture (2010); Reddy (2009b).

Note: Import penetration was calculated as a percentage of total domestic consumption.

A Closer Look at Pulse Imports, 2001–2012

Aggregate Imports

Overall, the importation of pulses has increased sharply between 2001 and 2012. During that period, imports rose from 0.6 million tons to 4.1 million tons, while the value imported rose from US$209 million to US$1.93 billion (in real terms). The compound annual growth rate in the volume of imports of agricultural commodities generally over that period was highest for edible oils (19.4 percent), followed by sugar (17.5 percent), and pulses (11 percent). In pulses, for the time period 2001–2003, the growth rate of import volumes was much higher than that of import value, indicating that growth in import prices was higher than growth in import volumes. Between 2004 and 2006 this trend somewhat reversed. During the final three years in the study period (2010–2012), the growth rates in both volume and value decreased by a similar order of magnitude (Figure 6.3).

Disaggregated Imports

Because there is a demand-supply gap for almost all types of pulses, almost all of them are imported into India to one degree or another. The major pulse imports include pea, pigeon pea, chickpea, black matpe, and lentil. Over the decade studied, significant variation is seen in the percentage shares of the different pulses out of total pulse imports. The share represented by peas increased significantly, while the shares of chickpea and pigeon pea declined. Figure 6.4 illustrates the distinctive pattern followed by peas (of all types) and yellow peas in particular.

Among all peas, yellow pea and split yellow pea had the biggest share in imports, and their quantities increased from 0.1 million tons in 2001 to 1.77 million tons in 2010—an almost 17-fold increase. The share of green pea was quite small. The percentage share of peas in total pulse imports increased from 17.7 percent in 2001 to almost 50 percent in 2010 (Figure 6.4). Likewise, when measured in terms of value, the increase has been very significant: from US$24.92 million in 2001 to US$389.5 million in 2010 (in 2005 constant US dollars). Figure 6.5 illustrates the distribution of the various pulses in total imports in 2001 and 2010. Chickpea, which had the highest share (24 percent) in 2001, underwent a serious decline to just 4 percent in 2010. For chickpea, the imports increased between 2001 and 2003, but experienced a significant fall between 2003 and 2011, with exceptions in 2005 and 2009 (Figure 6A.2 in the chapter appendix). There was a significant increase in chickpea imports in 2012, so its share in total imports reached 11 percent that

FIGURE 6.3 Compound annual growth rates (CAGR) for volume and value of pulses imports in India, 2001–2012

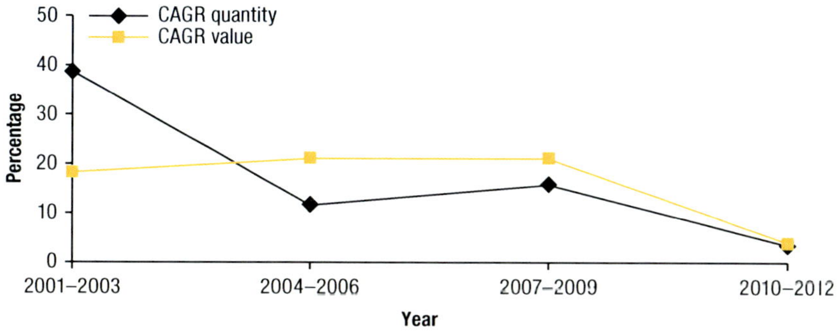

Source: Authors' calculations based on India Customs Data (2000–2012).

FIGURE 6.4 Increase in shares of all peas and yellow peas in total pulse imports, 2001–2012

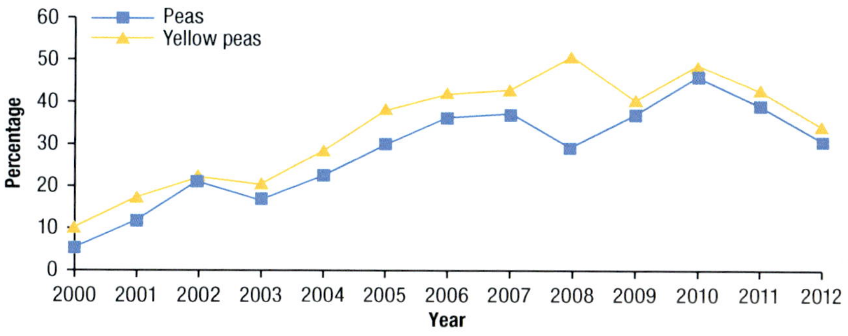

Source: Authors' calculations based on India Customs Data (2000–2012).

year. The movement of domestic prices and entry prices of imports is plotted in Figure 6A.1. Despite that increase, the overall trend in chickpea imports over the decade was downward.

One possible reason for the chickpea's declining share in imports could be this pulse's sustained increase in domestic production. Over the decade, chickpea production and yields in India increased by 12.4 percent and 5.1 percent, respectively (see Chapter 3 in this book). The falling import trend is consistent with a shift in consumption patterns over the decade, with consumption of chickpea decreasing by about 12.5 percent between 1988 and 2009. In contrast, pigeon pea imports exhibit a secular upward trend over the decade (Figure 6A.3 in the chapter appendix). The increase in pigeon pea imports was

FIGURE 6.5 Percentage of shares of different pulses in total pulses import, 2001 and 2010

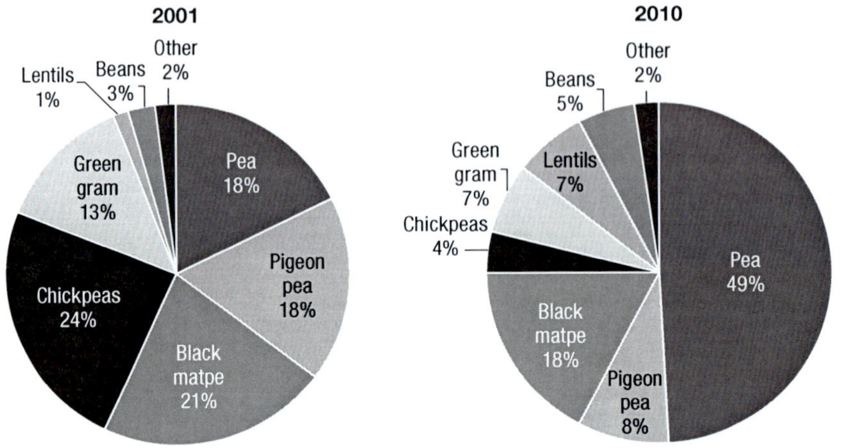

Source: Authors' calculations based on India Customs Data (2001–2012).

of a smaller order than the decrease in chickpea—namely, from 0.1 million tons in 2001 to almost 0.3 million tons in 2010. Despite the increase, the share of pigeon pea in total pulse imports declined sharply, from 17.7 percent in 2001 to 8.2 percent in 2010, implying that it was losing importance in the import basket. By 2012 the pigeon pea share had risen slightly to 9.6 percent of total pulse imports.

The pattern of pigeon pea imports is consistent with the domestic supply of pigeon pea over time. Both production and yield of pigeon pea increased, by 7.6 percent and 3.9 percent respectively, during the decade (see Chapter 3 in this book). This was preceded and accompanied by a 29.0 percent decline in its per capita consumption over the 21 years between 1988 and 2009. Among other pulses, the share of black matpe in the total pulses imports decreased from 21 percent in 2001 to 17 percent in 2010 and then declined further to 12 percent in 2012. Despite this, the trend in quantity imported has been positive, with imports of black matpe increasing from 0.1 million tons in 2001 to 0.60 million tons in 2010 (Figure 6A.4 in the chapter appendix). As with pigeon pea, the per capita consumption of black matpe also declined between 1988 and 2009—in this case by 25 percent

The shares of lentil and beans in total pulses imports increased and that of green gram decreased. During the decade, the total import quantity for beans showed an upward trend, whereas for green gram there has not been any particular pattern (Figure 6A.5 in the chapter appendix). The share of lentil in total quantity of pulses imported was just 1 percent in 2001, and it increased to 6 percent in 2010. The quantity of lentil imported was minimal until 2006, but after that year it began an upward trend, with an exceptional rise in 2012 (Figure 6A.4 in the chapter appendix), when it reached 17 percent of total pulse imports. Overall, between 1988 and 2009, the per capita consumption of lentil fell by 29.4 percent, and the per capita consumption of green gram fell by 37.5 percent.

To sum up, significant changes took place in pulse imports over the period studied, in volume, in value, and in composition. Chickpea and pigeon pea, which had dominated imports in the first half of the decade, were replaced in the second half by yellow pea. In the later part of the decade, lentil's share also increased significantly. Next, we take a closer look at the import trends of different types of pulse imports by studying their time-series properties.

Long-Run Dynamics among Different Types of Pulses Imports

In this section, we check for unit root and structural breaks to find out if there are mean reversal tendencies across the different pulse varieties imported. To trace the long-run dynamics between different types of pulse imports and co-movement across varieties as well as with domestic prices, we conduct cointegration tests. To determine the potential for imports to cool domestic markets, we formally study the role of one important pulse, pigeon pea, to examine whether and how much imports may be having this effect. For example, are the effects significant enough to bring down domestic prices, or are they merely stemming the rate of growth of prices? We assess the time it may take for the effects of imports (if we find any) to become manifest, since the time element would be very consequential.

Concerning co-movement, our hypothesis is that if two pulses are substitutes, an increase in the import of one should lead to a decrease in the domestic price or import of the other. By applying tests for cointegration, we assess whether there is any co-movement between the different types of pulses imported and their domestic prices. The multivariate cointegration method of Johansen (1988) and Johansen and Juselisus (1990) was employed to trace these long-run dynamics. The customs data we use, which are highly disaggregated at eight digits, enable us to look at the time-series behavior of each variety of imported pulse and the long-run dynamics between the varieties.

Pigeon pea and yellow pea are a good example of complementary varieties, since yellow pea is relatively cheaper than pigeon pea and is usually mixed with pigeon pea and sold in the market.

Because price changes are essentially a high-frequency phenomenon and agricultural trade and prices are highly impacted by seasonal factors, the high-frequency customs data used here are advantageous. They give the precise dates on which import consignments arrive. Thus, with the help of the dates listed in the dataset, we generate weekly imports of pulses by type (weekly frequency was chosen because import consignments for all pulses do not arrive daily). For domestic price data, however, we use daily frequencies, with information from the Center for Monitoring of the Indian Economy (CMIE). To match the domestic and import data, we average the daily prices over seven days to convert them to weekly domestic prices. According to Engle and Granger's (1987) formalization, two nonstationary series are said to be cointegrated if the following conditions are satisfied: (a) both the series are integrated and of the same order, and (b) there exists a linear combination of these series, which is I(0), i.e., stationary. Thus, while conducting cointegration analysis, the first step is to examine the integration properties of the relevant variable included in the model.

Stationarity

Following the modified Dickey-Fuller unit root tests, the log values of all the import quantity series and domestic price series are found to be nonstationary. However, the first difference of the series is integrated of order 0, meaning it is stationary. A second test using the method of Kwiatkowskii et al. (1989) is employed to ensure that the series are first-difference stationary. Taken together, the results suggest that all the series are first-difference stationary (Table 6.3). The unit root tests have been subjected to extensive theoretical and empirical research, which has shown them to be sensitive to the possibility of a structural break (Banerjee, Lumsdaine, and Stock 1992; Perron 1989; Zivot and Andrews 1992; Rappoport and Reichlin 1989). Thus we also conduct a Zivot and Andrews (1992) unit root test, accounting for one structural break in the time-series. We find that the results of the traditional unit root are validated even when we allow for a structural break in the series (see Table 6A.1 in the chapter appendix).

We selected a model for the different unit root tests by looking at the graph plot for import quantity and domestic prices of the different types of pulses. All the price series showed an upward trend, although no particular trend was

TABLE 6.3 Dickey Fuller Generalized Least Squares (DFGLS) and Kwiatkowski–Phillips–Schmidt–Shin (KPSS) unit root test results

	DFGLS		KPSS	
	Level	First difference	Level	First difference
Import quantity				
ln(Black_Matpe_Import)	−1.846	−3.465**	.259	0.002***
ln(Chickpea_Import)	−2.819	−3.951***	.164	0.003***
ln(Green_Gram_Import)	−2.682	−6.492***	.664	0.002***
ln(Lentil_Import)	−3.329	−4.605***	.514	0.004***
ln(Pigeon_Pea_Import)	−1.435	−4.973***	.167	0.002***
ln(Yellow_Pea_Import)	−1.530	−8.848***	1.410	0.002***
Domestic price				
ln(Black_Matpe_Price)	−1.571	−3.801***	2.320	0.035***
ln(Chickpea_Price)	−2.104	−3.689***	1.600	0.026***
Ln(Green_Gram_Price)	−1.735	−4.557***	1.360	0.054***
Ln(Lentil_Price)	−1.963	−17.559***	3.550	0.016***
Ln(Pigeon_Pea_Price)	−2.202	−8.341***	1.730	0.016***

Source: Authors' estimation.
Note: *** denotes 5 percent level of significance.

found in the import quantity series. Thus, while conducting the unit root test for different pulses, the domestic price series trend was included in the model. Since the different unit root tests assured us of the I(1) property, we next look for a cointegrating relationship to see if the different import quantities move in tandem over the long run. We test for pairwise co-movement between different pulse imports and their corresponding domestic prices series econometrically, in terms of their cointegration.

Cointegration

The Johansen and Juselius (1990) multivariate cointegration method involving up to k lags can be presented this way:

$$Y_t = \mu + A_1 Y_{t-1} + A_2 Y_{t-2} + \ldots\ldots\ldots\ldots\ldots + A_k Y_{t-k} + U_t,\ U_t \sim IN(0,\sigma) \qquad (1)$$

Where Y_t is a $(n \times 1)$ vector of variables, each of the A_i is a $(n \times n)$ matrix of parameters, μ is a vector of constants, and U_t is a $(n \times 1)$ normally and independently distributed vector of disturbances. If all the variables in Y_t are nonstationary and cointegrated, then equation 1 can be reformulated into a vector error correction model as follows:

$$\Delta Y_t = \Gamma 1 \Delta Y_{t-1} + \Pi Y_{t-1} + U_t \quad (2)$$

where Δ is the difference operator; $\Gamma_1........\Gamma_{k-1}$ are the coefficient matrices of short-term dynamics; and Π is a $(n \times n)$ matrix that contains information on long-run relationships. In fact, $\Pi = \alpha\beta'$ where α is a $(n \times k)$ matrix representing the speed of adjustment to disequilibrium and β defines the matrix of long-run coefficients.

Since the Johansen procedure is sensitive to changes in lag structure (Maddala and Kim 1998; Boswijk and Frances 1992), the optimal lag length for the model was decided based on the information criteria (Akaike Information Criteria [AIC], Schwarz Bayesian Criteria [SBC], and Hannan-Quinn Criteria, [HQC]).[5] Table 6A.4 in the chapter appendix presents the Johansen test results for the number of cointegrating vectors for different models. The various hypotheses tested—from no cointegration to increasing the number of cointegrating vectors—are reported in the first column of Table 6A.4. Table 6A.2 also reports the associated eigenvalues, trace statistics, and critical values.

We consider all the major pulses consumed in India, which include chickpea, pigeon pea, lentils, black matpe, and green gram. Although yellow pea constitutes a significant share, it could not be included in the analysis since it has no domestic production and thus no domestic price for it is available. Two main pulses, chickpea, and pigeon pea, are paired with other pulse imports and their domestic prices to examine whether they act as substitutes for one another. For instance, to check the long-run interdependence between chickpea and pigeon pea, chickpea, and pigeon pea imports and their corresponding domestic prices are included in the model. Results from Table 6A.2 show that both chickpea and pigeon pea are indeed pairwise cointegrated with different pulses imports and their domestic prices. At the maximum, two cointegrating relations were found for the different models.

Chickpea import's effect on other pulses. Based on the cointegration relation, the long-run association between chickpea and other pulses can be represented by equations 3 through 6:

$$Chickpea\_Import_t = 11.82 Pigeonpea\_Import_t - 3.75 Pigeonpea\_Price_t - 17.50 Chickpea\_Price_t \quad (3)$$

5 Information criteria suggesting the minimum lag was selected.

$$Chickpea\_Import_t = 2.22 Greengram\_Import_t - 3.95 Greengram\_Price_t - 2.07 Chickpea\_Price_t \quad (4)$$

$$Chickpea\_Import_t = 8.99 Blackmatpe\_Import_t - 15.21 Chickpea\_Price_t \quad (5)$$

$$Chickpea\_Import_t = -.147 Lentil\_Import_t + 1.66 Chickpea\_Price_t \quad (6)$$

This means an increase in chickpea imports has a negative effect on the domestic price of pigeon pea and green gram, but it has a positive effect on their import quantity. For instance, a 1 percent increase in chickpea imports is associated with a 4 percent decrease in pigeon pea's domestic price and a 17 percent decrease in chickpea's domestic price. However, it is positively associated with pigeon pea's import quantity. The same 1 percent increase in chickpea imports is also associated with a 4 percent reduction in green gram's domestic price and a 3 percent reduction in chickpea's domestic price, but once again it has a positive association with green gram's import quantity. Chickpea imports are also associated positively with black matpe imports and negatively with the black matpe domestic price, although in this case the price coefficient is insignificant. Chickpea import is negatively associated only with the import of lentils. Thus from these results, we can conclude that there is some co-movement between imports of different types of pulses and the domestic prices of pulses and imports.

Pigeon pea import's effect on other pulses. The long-run association between pigeon pea and other pulses is represented by equations 7 through 10:

$$Pigeonpea\_Import_t = -.38 Greengram\_Import_t + 1.36 Greengram\_Price_t + 1.06 Pigeonpea\_Price_t \quad (7)$$

$$Pigeonpea\_Import_t = .15 Lentil\_Import_t + .90 LentilPrice_t + .72 Pigeonpea\_Price_t \quad (8)$$

$$Pigeonpea\_Import_t = 2.69 Blackmatpe\_Import_t - .31 Blackmatpe\_Price_{Pt} - 2.48 Pigeonpea\_Price_t \quad (9)$$

$$Pigeonpea\_Import_t = 0.08 Chickpea\_Import_t + 1.4 Chickpea\_Price_t + .31 Pigeonpea\_Price_t \quad (10)$$

Pigeon pea's import is positively associated with the import quantities of lentils, black matpe, and chickpea. Only green gram's import is negatively associated with pigeon pea import. Regarding price associations, the increased import of pigeon pea is negatively associated only with the domestic price of black matpe, pointing to the possibility of a substitution effect between these two. Since for several pulses there are no clear links in prices or in import quantities, as expected the trade data reveal only some substitution effect between different types of pulses that tie imports closely to the demand-supply gap by type.

Evolution of India's Trading Partnerships in Pulses

Trade evolves at two margins. The *intensive* margin is where increases or decreases occur in exports or imports with existing trading partners or commodities, while the *extensive* margin is where new trading partners and new varieties emerge. Having looked at the different facets of India's pulse imports on the intensive margin, next, we look at the dynamics of trade on the extensive margin. After 2001, pulse imports first expanded on the extensive margin and then expanded at the intensive margin. The initial turnover occurred as several countries experimented with the Indian market just as imports commenced. Significant imports of pulses started around 1998–2000, mainly from Australia, Canada, Myanmar, Tanzania, Thailand, Turkey, the United States, and Uzbekistan. Over time, the volume of pulse imports increased, and India also started importing from additional countries. The newer trading partners included China, Ethiopia, Indonesia, Malawi, Mozambique, and the Russian Federation. Imports from the latter group of countries, however, remained on a small scale over the decade, and excluding Malawi, Mozambique, and the Russian Federation, their shares remained small.

Another pattern is that for each pulse there has usually been a single partner with a significant market share, generally greater than 50 percent and in some cases reaching 80 percent. Examples of such key leaders at different times include Australia for chickpea, Canada for pea and lentil, and Myanmar for pigeon pea, green gram, and black matpe. Figure 6.6 and Table 6.4 present the evolution of India's trade with these countries over time; in Figure 6.6 Panel A shows the percentage shares of different countries in India's total pulse imports for 2001, and Panel B shows the same for 2010. While Myanmar's share of total Indian imports was the largest share in 2001, at 41 percent, by 2010 its share had fallen to 21 percent and by 2012 to 17 percent. Still, Myanmar remained the top exporter of green gram, pigeon pea, and black matpe to India. Canada's share increased from 21 percent in 2001 to 32 percent in 2010 but then fell to 28 percent in 2012. During the first half of the decade, beginning in 2000, Canada was exporting chickpea, lentil, and pea to India, but in the second half of the decade its chickpea

FIGURE 6.6 Percentage shares of different countries in total pulses imports by India, panel A (2001) and panel B (2010)

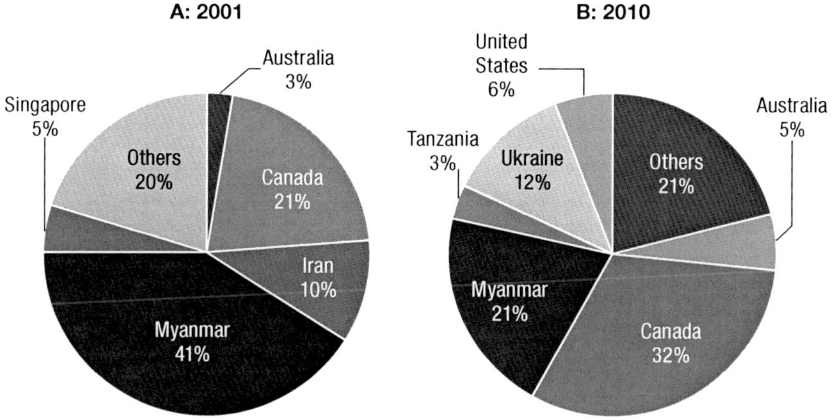

Source: Authors' calculations based on India Customs Data (2001–2012).

TABLE 6.4 Percentage of shares of different countries in total pulses imports by India, 2001–2012

Country	2001	2005	2010	2012
Australia	3	3	5	10
Canada	21	40	32	28
Iran	10	—	—	—
Myanmar	41	36	21	17
Russian Federation				13
Singapore	5	—	—	—
Tanzania	—	4	3	2
Ukraine			12	1
United States	—	3	6	3
Others	20	14	21	23

Source: FAOstat, Food and Agriculture Organization (2012).
Note: — = data not available.

exports became almost negligible and only lentil and yellow pea were exported to India. Australia's share remained more or less the same between 2001 and 2005, but then it increased, reaching 10 percent in 2012. The Russian Federation is the latest addition to the list of trading partners, with its share of 13 percent in pulses imports by India in 2012.

Imports of Major Pulses, Disaggregated by Type

Peas

Peas being the most important import, their interplay on both the extensive and intensive margins deserves special attention. Peas are imported mainly from Canada, the United States, Australia, and the Russian Federation, in that order. Although this has varied over the years, on average 70 percent to 80 percent of pea imports have come from Canada, with around 10 percent coming from the United States and the rest from other countries. Canadian imports have been consistent and have trended upward over the decade studied, except for sudden drops in 2008, 2011, and 2012. In those exceptional years, the share of US imports increased significantly, but despite this, the US share in India's pea imports was quite small when compared with Canadian imports in all the other years (Figure 6.7). In fact, Canada is the world's largest producer and exporter of peas (FAO 2012). A significant share of its production is exported to India. Of 3.06 million tons of peas produced in Canada in 2010, 2.79 million tons were exported, 44 percent of which landed in India. After India, the most important destinations for Canadian peas have been China and Bangladesh, although their shares have been very small in comparison with India's.

Another factor behind the dominance of Canadian pea imports in India is that Canadian peas tend to be cheaper than competitors'. Based on the customs data, the unit value of Canadian peas in 2010 was US$294 per ton, as compared with a much higher value of US peas (US$421 per ton) (Table 6.5). However, during 2009–2012 the unit value of US peas also came down, reaching near parity with Canadian peas, which could explain the decline in Canada's share of pea imports and the rise in the US share after 2010. Based on the customs data, most of the imported peas land at Mumbai port. Shipments from Canada take around 35 days to reach Mumbai port, while those from the United States take around 45 to 48 days, so shipping costs could be another reason for the higher demand for Canadian peas in India.

Chickpea

India's main trading partners in chickpea are Australia, Myanmar, Canada, and Tanzania. Unlike pea imports, chickpea imports expanded more on the intensive margin. Until 2005 a significant quantity of chickpea was being imported from Canada, Myanmar, and Tanzania, but after that year Australia has been the main exporter of chickpea to India (Figure 6.8). The possible reasons for this change are a decrease in chickpea production in Canada and an

FIGURE 6.7 Major trading partners of India in peas, 2001–2013

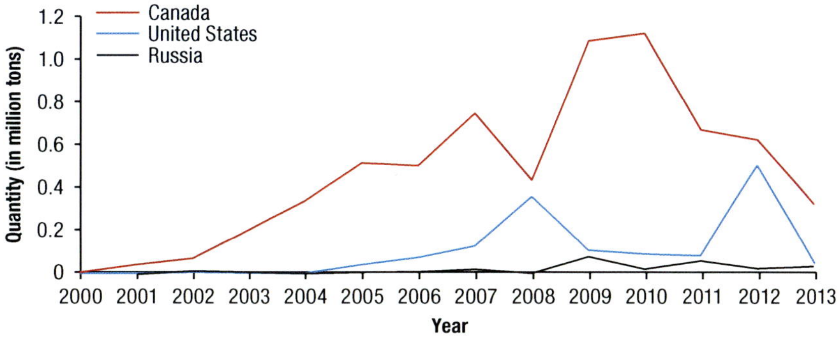

Source: Authors' calculations based on India Customs Data (2001–2013).

TABLE 6.5 Triennium average value of peas imported by India from Canada and the United States (US$ per ton)

Triennium	Canada	United States
2001–2003	212	272
2004–2006	246	286
2007–2009	373	256
2010–2012	364	304

Source: Authors' calculations based on India Customs Data (2001–2012).
Note: Values are in constant 2005 US$.

increase in the price of Myanmar's chickpea (FAO 2012).[6] In the second half of the decade, Australia's share of imports rose. The Russian Federation's share in total chickpea imports has been low (approximately 2 percent) throughout the decade, except in 2012 when it accounted for a peak 17 percent share.

Furthermore, from 2001 to 2003 the average unit value of chickpea imported from Myanmar was US$66 per ton, which was very much lower than the average unit value of Australian chickpea (US$613 per ton). However, after 2005 the unit value of Myanmar's chickpea increased significantly; for example, it was US$550 to US$600 per ton during the 2007–2009 triennium and remained high even during the 2010–2012 triennium (Table 6.6). In terms of the time taken to landing ports in India, there might be some advantage for imports from Myanmar, as it takes around

6 CAGR for chickpea production in Canada between 2000 and 2012 was −0.072.

TABLE 6.6 Triennium average unit values of imported chickpea from Australia and Myanmar, 2001–2012 (US$ per ton)

Triennium	Australia	Myanmar
2001–2003	613	66[a]
2004–2006	415	436
2007–2009	475	557
2010–2012	539	830

Source: Authors' calculations based on India Customs Data (2001–2012).

Note: Values are in constant 2005 US$. [a] The low price of Myanmar chickpea might be due to its low volume as well as to its desi variety, which is different from the one imported from other countries.

FIGURE 6.8 Major trading partners of India in chickpea, 2000–2012

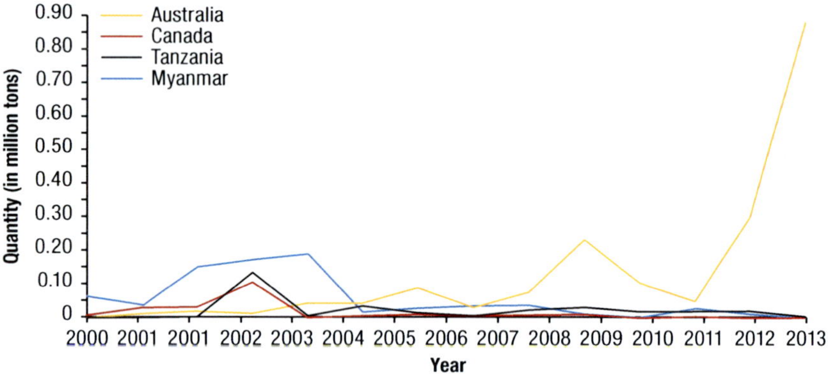

Source: Authors' calculations based on India Customs Data (2001–2012).

26 days for a shipment to reach India from Australia and only 9 days to do so from Myanmar. Note that Myanmar and India have the same seasons for production of pulses, so exploiting price differences across seasons would require incurring storage costs, which can be high due to susceptibility to insect damage.

Lentil

As in the case of peas, Canada is also the world's largest producer and exporter of lentils, with about 2 million tons of annual production, of which around

TABLE 6.7 Triennium average unit value of lentil imported by India, 2001–2012 (US$ per ton)

Triennium	Canada	Australia	United States
2001–2003	308	381	431
2004–2006	364	435	298
2007–2009	711	658	390
2010–2012	543	547	637

Source: Authors' calculations based on India Customs Data (2001–2012).
Note: Values are in constant 2005 US$.

FIGURE 6.9 Major trading partners of India in lentil, 2000–2012

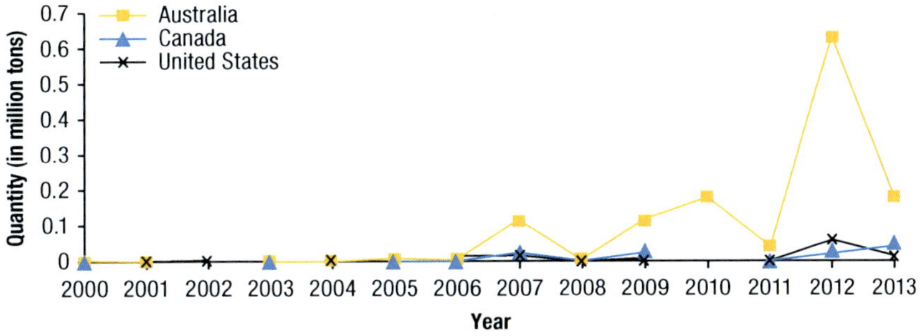

Source: Authors' calculations based on India Customs Data (2001–2012).

80 percent is exported. In the global lentil market, after Canada comes Australia, United States, and Turkey. Canada is also the major exporter of lentil to India, accounting for 75 percent to 85 percent of India's import total, with the rest being imported mainly from Australia and the United States. Lentil imports in India picked up after 2006. In terms of unit values, the costliest lentil seems to be that exported from Turkey. For instance, the unit values (in US$ per ton) for lentil exported from Canada, Australia, United States, and Turkey in 2011 were 761, 677, 712, and 945, respectively (FAO 2012). A comparison of three-year average unit values of lentil imported from different trading partners reveals that despite the availability of lentil at low price from the United States (in 2004–2006 and 2007–2009) and Australia (2007–2009), the import was highest from Canada (Table 6.7). Australia's share in total imports showed a significant increase in 2012 (Figure 6.6).

Pigeon Pea

India imports pigeon pea largely from Myanmar, Malawi, Mozambique, and Tanzania, and to a smaller degree from Australia, the United States, Canada, Ethiopia, and China. Compared to the first half of the 2000 decade, when Myanmar made up around 70 percent to 80 percent of India's pigeon pea imports, in the second half of the decade Myanmar's share receded to 50 percent. At the same time, the share of imports made up by Tanzania, Mozambique, and Malawi increased from 1 percent to 5 percent in the first half to 15 percent to 20 percent in the second half of the decade (Figure 6.10).

The production of pigeon pea in Myanmar shows an upward trend over time, although the country's compound annual growth rate of production between 2000 and 2012 was just 0.14 percent. Of its total production, Myanmar exported around 20 percent to 30 percent to India. Table 6.8 shows the three-year average unit values of pigeon pea imported by India from major trading partners. It becomes clear from the table that price could not be the important factor for the high demand for Myanmar pigeon pea, as there were few years when the unit value of Myanmar pigeon pea was lower than that imported from other countries. Distance could be another reason; compared to Tanzania (14 days) and Mozambique (15 days), a shipment from Myanmar takes much less time (9 days) to reach India.

Green Gram

From the beginning of the decade studied, green gram was being imported into India mainly from Myanmar. Other trading partners include China, Australia, Tanzania, and Uzbekistan, in that order. However, the shares of these countries have been quite small, although the share of Myanmar itself fluctuated throughout the decade. The trade spikes in Myanmar's exports in 2002, 2006, 2009, and 2012 (Figure 6.11) were due to these being years when green gram production in India was significantly lower than normal. At the same time, Myanmar's pulse production rose throughout the decade. Myanmar's share in India's total pulses imports was around 60 percent to 65 percent until 2005; it increased to 80 percent to 85 percent between 2006 and 2007, then declined to 40 percent to 42 percent between 2008 and 2012. Australia's share increased from 11 percent to 21 percent between 2010 and 2012.

The unit values of green gram imported from different countries tend to be quite similar. In fact, in some time periods, the unit value of imports was

FIGURE 6.10 Major trading partners of India in pigeon pea, 2000–2012

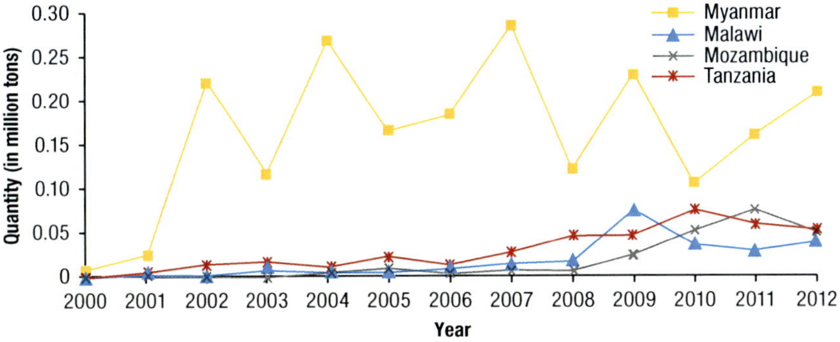

Source: Authors' calculations based on India Customs Data (2001–2012).

TABLE 6.8 Triennium average unit value of pigeon pea imported by India, 2001–2012 (US$ per ton)

Triennium	Myanmar	Malawi	Mozambique	Tanzania
2001–2003	309	232	279	493
2004–2006	306	409	364	631
2007–2009	510	698	641	741
2010–2012	773	687	605	679

Source: Authors' calculations based on India Customs Data (2001–2012).
Note: Values are in constant 2005 US$.

higher from Myanmar than from other countries (Table 6.9). High production, easy availability, and shorter shipping time could all be possible reasons for Myanmar retaining the lion's share of green gram imports.

Black Matpe

Myanmar is the major exporter to India of black matpe, accounting for a share of more than 80 percent of India's import throughout the decade. Only a very small quantity of black matpe was imported from Malaysia. Other countries that exported black matpe to India were Ethiopia, Malawi, and Thailand, but their shares were extremely small (Figure 6.12). Note that Singapore figures in among exporters because exports from other countries such as Myanmar are often channeled through Singapore.

FIGURE 6.11 Trends in imports of green gram from major trading partners, 2000–2012

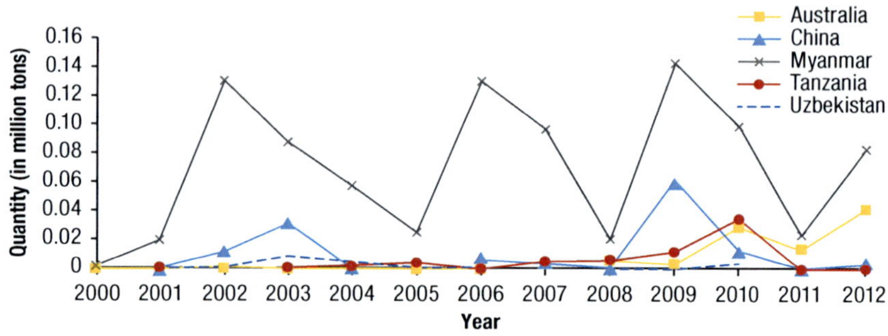

Source: Authors' calculations based on India Customs Data (2001–2012).

TABLE 6.9 Triennium average unit value of green gram imported by India, 2001–2012 (US$ per ton)

Triennium	Australia	China	Myanmar	Tanzania
2001–2003	436	339	386	406
2004–2006	496	534	502	443
2007–2009	1,238	785	744	606
2010–2012	995	1126	948	—

Source: Authors' calculations based on India Customs Data (2001–2012).
Note: Values are in constant 2005 US$. — = data not available.

FIGURE 6.12 Major trading partners of India in black matpe, 2000–2012

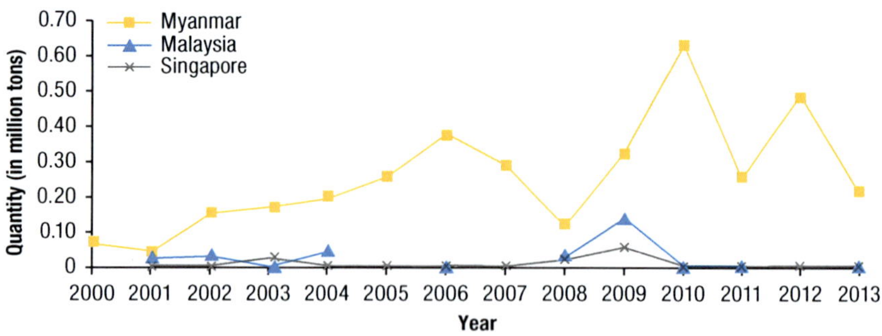

Source: Authors' calculations based on India Customs Data (2001–2012).

The Cooling Effect of Imports on Prices: The Case of Pigeon Pea

Having looked at the dynamics of the pulse trade in detail, we now explore, as a case study, the import of pigeon pea by India and its effects on domestic prices.[7] In one of the earliest studies done on cooling effects in Indian agricultural imports, Mann (1967) looked at the importation of cereals under Public Law 480 and its effects on domestic prices in India. The findings in Mann's study show that the imports led to lower prices and a decline in domestic supply, but also that the decrease in domestic supply was less than the quantity imported. Thus there was a net addition to the quantity available for consumption, which is a significant contribution in a shortage economy.

An alternative way of investigating the impact of imports on domestic prices is by looking at price transmission from international markets to domestic markets. A large number of studies have examined the degree of price transmission between markets within a country and to the rest of the world (among others, see Abdulai 2000 for Ghana and Negassa and Myers 2007 for Ethiopia). Conforti (2004) finds price transmission occurring in 16 countries: in Ethiopia for wheat, sorghum, and maize; in Ghana for wheat; and in Senegal for rice. There are indeed also cases where price transmission is weak (for example, for maize in Africa), implying that a cooling effect would be less likely in such cases.

Evidence also exists, however, for the converse of a cooling effect of imports on prices. The food crisis of 2007–2008, for example, provoked several export restrictions in countries around the world, mainly for staple foods. For example, Malawi, Tanzania, and Zambia banned the export of maize, while India banned the export of rice. Minot (2011) argues that these export bans probably contributed to rising grain prices in Africa south of the Sahara more so than in landlocked countries among them (Staatz et al. 2008). Although India liberalized trade in the 1990s, a number of major food commodities still do not have an open trade regime, and this includes pulses. Imports are restricted through tariff and nontariff barriers. The government of India maintains these restrictive import policies in a bid to protect the domestic agricultural sector, but only for sectors where there is significant domestic production. Consider, for example, the case of edible oils: at present, most of the domestic demand for edible oils in India is met through imports, and the trade regime for these oils is liberal.

---

7 We also present the domestic price movement of different varieties of pulses in some major markets in 2014 and 2015 in Table 6A.3 in the chapter appendix.

This case study of pigeon pea could, therefore, be useful to the government in better understanding the extent and nature of the roles trade can play in price management. In commodities that have high demand but persistently inadequate supply, more free trade could help, although there are significant qualifications to keep in mind. First, it is possible that in some cases there might not be adequate supply in the international market, as is the case in pulses. Second, for imports to affect prices, there is likely to be a time lag, implying that advance planning is needed. Finally, differences in production seasons might not always be well aligned.

Time-Series Analysis for Cooling Effect

The two relevant series—pigeon pea weekly imports and domestic prices—are found to be integrated of order 1. A vector error correction model (VECM) is then employed to study the dynamics of pigeon pea imports and Wholesale Price Index (WPI). This model allows us to delineate the short-run from the long-run import-price dynamics. The impulse-response graphs help us visually interpret the short-term adjustment by imports in response to a unitary shock in prices and vice versa. The cointegrating equation provides an estimate of the long-run import-price equlibrium relationship, and the adjustment coefficients describe the speed of adjustment of the system toward correcting the previous period's equilibrium error in the subsequent period.

Since both log (Imports) and log (WPI) series are integrated of the same order [I(1)], the existence of a cointegrating relationship can be tested for and represented by an error-correction mechanism in case cointegration is found (Engle and Granger 1987). For the current bivariate system of real imports and prices, the cointegrating rank can at most be 1. If such is the case, the two series are expected to maintain a long-run equilibriating relationship, which is given by the cointegrating equation, and the changes in each variable in the short-run can be modeled through an error correction representation. The test statistics for the cointegrating equation indicated that the long-run relationship between prices and imports is significant. Moreover, the coefficient for imports in the equation is negative and statistically significant, implying that imports and prices share an inverse relationship. Prices are highly elastic to changes in imports, as 1 percent increase in imports is estimated to bring down prices by as much as 3 percent in the long run.

Table 6.10 presents the results from the vector error correction model, where the speed of adjustment coefficients indicate a very sluggish error adjustment by the model. When predictions from the cointegrating equation

TABLE 6.10 Results from the vector error correction model (VECM)

Variable	Δ ln (WPI)	Δ ln (Import)
Speed of adjustment	−0.0076***	0.130***
	(0.0027)	(0.0227)
Δ ln (WPI): lag 1	0.0663	0.708*
	(0.0444)	(0.3689)
Δ ln (WPI): lag 2	0.0732	−0.155
	(0.0464)	(0.3850)
Δ ln (WPI): lag 3	−0.0600	−0.115
	(0.0464)	(0.3852)
Δ ln (Import): lag 1	−0.0129*	−0.465***
	(0.0073)	(0.0610)
Δ ln (Import): lag 2	−0.00903	−0.386***
	(0.0065)	(0.0542)
Δ ln (Import): lag 3	−0.0126**	−0.306***
	(0.0052)	(0.0428)
Constant	0.00132	0.0000766
	(0.0009)	(0.0072)
Number of observations	518	
AIC	−5.84	
HQIC	−5.782	
SBIC	−5.697	
log likelihood	1528.65	
Chi2	22.77	392.6
R2	0.043	0.435
Root Mean Square Error (RMSE)	0.019	0.162

Source: Authors' calculations.

Note: ***p < 0.01, **p < 0.05, and *p < 0.1. The figures within the parantheses are standard errors.

are positive, prices are above their equilibrium value. When there is such an equilibrium imbalance, the error adjustment for prices is 0.01 in the current period, which means that prices fall back toward imports, although by a negligible proportion, to correct the error in the past period. In comparison, imports, having a coefficient of 0.13, display a somewhat rapid adjustment mechanism of moving up toward prices while the prices are adjusting. This could be the case because imports are comparatively variable relative to prices, which tend to be sticky in the short run. Also, since the data on imports and prices are considered at a weekly frequency, imports are more volatile compared to prices. Nevertheless, both the adjustment coefficients are significant at the 1 percent level of significance.

Short-run Dynamics of Imports and Prices

The short-term dynamics between the variables in the import-price model are demonstrated by generating the impulse response functions (IRFs). Figure 6.13 displays the two IRFs, which forecast the time paths of a variable when exposed to a one-standard-deviation innovation in the other endogenous variable. We find that a unitary shock in imports stablizes prices only after 20 weeks. In the interim, prices of pigeon pea continue to rise but, importantly, they rise with a concave trajectory—that is, the rate of increase in prices falls as weeks go by. In both the cases, a unitary shock causes permanent innovations in the time paths of the influenced variables.

Granger Causality Tests

To assess a causal link in a time-series analysis, Granger causality tests are a useful diagnostic tool to check the direction of causality, particularly in situations where there is no theoretical justification for the direction of causality. In a bivariate setup, a variable X is said to "granger-cause" variable Y if controlling for the past values of Y, the lagged values of variable X are useful in predicting Y. In our case, both the null hypothesis of imports not granger-causing prices and prices not granger-causing imports are rejected (Table 6.11).

Seasonality and Imports

One aspect of imports having a cooling effect on prices relates to agricultural seasonality, which includes pulse cultivation. Indian agricultural markets are characterized by wide variation in prices that are mainly driven by fluctuations in market arrivals. The nature of supply and demand for agricultural products creates instability in prices for producers and consumers as well as instability in farmers' income. About a dozen varieties of pulses are grown in India, with pigeon pea among the most widely consumed. Market arrivals data for Karnataka, for example, show that seasonality is quite important in pigeon pea markets, with the highest-volume arrivals occurring from January to March. At the same time, the lowest-volume arrivals are observed in the

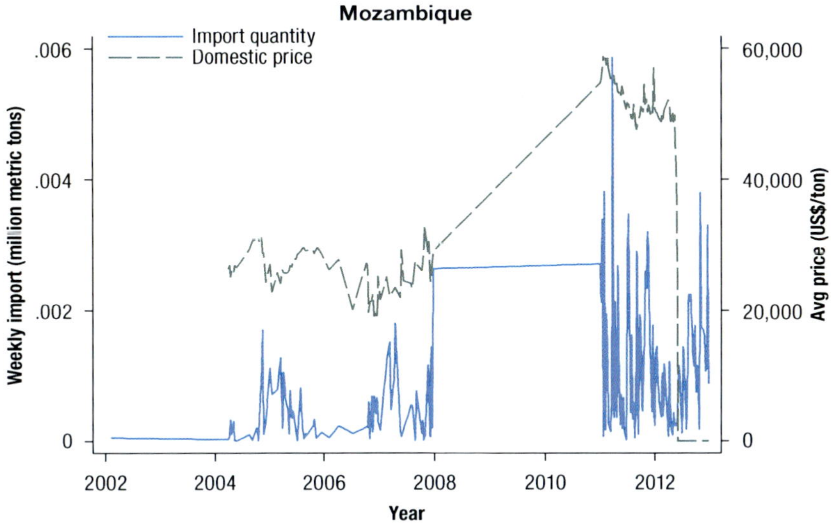

Source: Indian Customs Data; data on wholesale prices; and India, Ministry of Finance.

policy needs to become compatible with this reality. At a minimum, policy needs to be consistent over time. Generically, trade policy in India, especially for agricultural commodities, has been quite inconsistent, with both import and export policies for pulses changing over time. Frequently changing trade policies bring uncertainty both in the market and among traders. A stable trade policy would ensure consistent supply in the domestic market and, possibly, better prices.

India is among the world's top importers of pulses. Having such a large share means India's import demand can have a bearing on world prices. Furthermore, to ensure a consistent supply of pulses from the international market, diversification might be needed across trading partners and pulse varieties. The analysis shows that Myanmar, Canada, and Australia have been the most important sources of pulse supply to India. For each variety of pulses, there are just a few major exporting partners that cover more than 50 percent of the supply, and in some cases, a few countries account for more than 80 percent of the supply. This is problematic because countries with such big shares can easily alter the supply and the prices, and production shocks in these countries can affect the availability of pulses in India. Thus there is a need to diversify across different countries to fulfill India's import demand. Over the last five or six years of the study (that is, 2008–2012), imports expanded on the extensive margin as well. This could mitigate concerns that relying on external markets could be detrimental if production shocks occurred in the exporting country.

Looking at the market cooling effects of pulses, we find that a unitary shock in imports at first leads to a sustained increase in prices lasting up to 20 weeks, after which the prices stabilize. Thus imports do affect prices, but in the form of moderating their rate of increase rather than bringing them down. A few policy lessons follow from this finding. First, imports need to be operationalized quickly, since it takes quite some time for imports to bear on prices. Second, the size of the imports needs to be increased, because currently they are not sufficient to bring down prices but only sufficient to moderate the rate of price growth.

References

Abdulai, A. 2000. "Spatial Price Transmission and Asymmetry in the Ghanaian Maize Market." *Journal of Development Economics* 63: 327–349.

Agbola, W. F. 2003. "Pulse Policy Reform in India: Implications for Trade, Prices, and Production, 1970–1999." Paper presented at the 47th annual conference of the Australian Agricultural and Resource Economics Society, Fremantle, Australia, February 11–14.

Banerjee, A., R. L. Lumsdaine, and J. H. Stock. 1992. "Recursive and Sequential Tests for a Unit Root: Theory and International Evidence." *Journal of Business Statistics* 10: 271–287.

Boswijk, H. P., and H. Frances. 1992. "Dynamic Specification and Cointegration." *Oxford Bulletin of Economics and Statistics* 54: 369–381.

Conforti, P. 2004. *Price Transmission in Selected Agricultural Markets.* Commodity and Trade Policy Research Working Paper 7. Rome: Food and Agriculture Organization of United Nations (FAO).

Engle, R. F., and W. J. Granger. 1987. "Cointegration and Error Correction: Representation, Estimation, and Testing." *Econometrica* 55: 251–276.

FAO (Food and Agriculture Organization of the United Nations). 2012. FAOSTAT database. Accessed September and October 2013, 2014. http://faostat.fao.org.

Gokarn, S. 2011. "The Price of Protein." *Macroeconomic and Finance in Emerging Market Economies* 4 (2): 327–335.

IIPR (Indian Institute of Pulses Research). 2013. *Annual Report 2012–13.* Kanpur. India: IIPR. Accessed June 2016. www.iipr.res.in/pdf/annual_english_18july2013.pdf.

India, Ministry of Agriculture. 2010. *Agricultural Statistics at a Glance 2010*. New Delhi: Oxford University Press; Department of Agriculture and Cooperation, Directorate of Economics and Statistics.

———. 2013. *Agricultural Statistics at a Glance 2013*. New Delhi: Indian Council of Agricultural Research Accessed August 2016. http://agricoop.nic.in/agristatistics.htm.

India Customs Data. 2000–2012. Trade Genius—Export Import Database. Accessed in 2013, 2014. www.tradegeniusindia.com/.

Johansen, S. 1988. "Statistical Analysis of Cointegration Vectors." *Journal of Economic Dynamics and Control* 12 (2–3): 231–254.

Johansen, S., and K. Juselius. 1990. "Maximum Likelihood Estimation and Inference on Cointegration—with Applications to the Demand for Money." *Oxford Bulletin of Economics and Statistics* 52 (2): 169–210.

Kumar, P., P. Shinoj, S. S. Raju, A. Kumar, K. M. Rich, and S. Msangi. 2010. "Factor Demand, Output Supply Elasticities and Supply Projections for Major Crops of India." *Agricultural Economics Research Review* 23: 1–14.

Kwiatkowski, D., P. C. B. Phillips, P. Schmidt, and Y. Shin. 1989. "Testing the Null Hypothesis of Stationarity against the Alternative of a Unit Root." *Journal of Econometrics* 54: 159–178.

Maddala, G. S., and I. M. Kim. 1998. *Unit Roots, Cointegration, and Structural Change*. Oxford, UK: Oxford University Press.

Mann, J. S. 1967. "The Impact of Public Law 480 Imports on Prices and Domestic Supply of Cereals in India." *American Journal of Farm Economics* 49 (February): 131–146.

Mehta, R., and P. K. Srivastava. 2000. "Analysis of Seasonality of Prices of Agricultural Commodities." *Agricultural Situation in India* 57 (6): 311–317.

Minot, N. 2011. *Transmission of World Food Price Changes to Markets in Sub-Saharan Africa*. IFPRI Discussion Paper 1059. Washington, DC: International Food Policy Research Institute.

Mueller R. A. E., P. Parthasarathy Rao, and K. V. Subba Rao. 1990. "Pigeonpea: Markets and Outlook." In *The Pigeonpea*, edited by Y. L. Nene, S. D. Hall, and V. K. Sheila, 457–479. London, UK: CAB International; Patancheru, India: International Crops Research Institute for Semi-Arid Tropics (ICRISAT).

Murugkar, M. 2006. *Reforms in Food Corporation of India: Case of Wheat Imports*. Briefing Paper Granite 1/2006. Jaipur, India: CUTS International.

Negassa, A., and R. Myers. 2007. "Estimating Policy Effects on Spatial Market Efficiency: An Extension to the Parity Bounds Model." *American Journal of Agricultural Economics* 89: 338–352.

Perron, P. 1989. “The Great Crash, the Oil Price Shock, and the Unit Root Hypothesis.” *Econometrica* 57: 1361–1401.

Price, G. K., R. Landes, and A. Govindan. 2003. “India’s Pulses Sector: Results of Field Research.” Report from the Economic Research Service, US Department of Agriculture, Washington, DC.

Rappoport, P., and L. Reichlin. 1989. “Segmented Trends and Non-stationary Time Series.” *Economic Journal* 99: 168–177.

Reddy, A. A. 2009a. “Policy Options for India’s Edible Oil Complex.” *Economic and Political Weekly* 44: 22–24.

———. 2009b. “Pulses Production Technology: Status and Way Forward.” *Economic and Political Weekly* 44: 73–80.

Sathe, D., and S. Agarwal. 2004. “Liberalization of Pulses Sector: Production, Prices, and Imports.” *Economic and Political Weekly* 39: 3391–3397.

Staatz, J., N. Dembélé, V. Kelly, and R. Adjao. 2008. “Agricultural Globalization in Reverse: The Impact of the Food Crisis in West Africa.” Background paper for the Geneva Trade and Development Forum, September 17–20. Crans-Montana, Switzerland. Department of Agricultural, Food, and Resource Economics, Michigan State University, East Lansing, MI, US. Accessed July 2016. www.aec.msu.edu/fs2/srai/Agricultural_Globalization_in_Reverse_MSU_CransMontana_paper_final.pdf.

UN Comtrade database. Accessed March 2016. https://comtrade.un.org/.

USAID. 2012. “The Indian Market for Pigeon Pea.” Kenya Horticulture Competitiveness Project (KHCP). Accessed June 2016. http://pdf.usaid.gov/pdf_docs/PA00KC9G.pdf.

Von Oppen, M. 1981. “Marketing of Pigeonpea in India.” In *Proceedings of the International Workshop on Pigeonpeas*, Vol. 1: 332–343. Workshop held December 15–19, 1980. Patancheru, India: International Crops Research Institute for Semi-Arid Tropics (ICRISAT).

Walker, T., S. Silim, B. Cunguara, C. Donovan, P. R. Parthasarathy, M. Amane, and M. Siambi. 2015. “Pigeonpea in Mozambique: An Emerging Success Story of Crop Expansion in Smallholder Agriculture.” Working Paper 78E. Ministry of Agriculture and Food Security, Directorate of Economics, Maputo.

Zivot, E., and D. W. K. Andrews. 1992. “Further Evidence on the Great Crash and the Oil Price Shock.” *Journal of Business and Economic Statistics* 10 (3, July): 251–270.

Appendix

TABLE 6A.1 Zivot Andrews unit root test for structural break

Variable	Level	First difference
Import quantity		
ln(Bm_Im)	−4.02	−23.03***
ln(Cp_Im)	−5.22	−16.47***
ln(Gg_Im)	−5.44	−18.84***
ln(L_Im)	−4.117	−24.03***
ln(Pp_Im)	−5.23	−17.152***
ln(Yp_Im)	−3.031	−17.156***
Domestic prices		
ln(Bm_P)	−4.119	−35.82***
ln(Cp_P)	−2.443	−35.68***
Ln(Gg_P)	−3.012	−39.14***
Ln(L_P)	−4.401	−33.39***
Ln(PP_P)	−3.260	−36.64***

Source: Authors' estimation.

Note: *** shows 1 percent level of significance.

TABLE 6A.2 Johansen trace test statistics to determine the number of cointegrating rank

Chickpea and pigeon pea			
Maximum rank	Eigenvalue	Trace statistics	5% critical value
0	—	310.66	47.21
1	0.125	121.51	29.68
2	0.072	10.02*	15.41
3	0.006	1.30	3.76
4	0.002		

Chickpea and green gram			
Maximum rank	Eigenvalue	Trace statistics	5% critical value
0	—	254.37	47.21
1	0.207	125.62	29.68
2	0.194	6.20*	15.41
3	0.009	0.67	3.76
4	0.001		

Chickpea and black matpe			
Maximum rank	Eigenvalue	Trace statistics	5% critical value
0	—	375.35	47.21
1	0.349	137.73	29.68
2	0.208	8.12*	15.41
3	0.013	0.68	3.76
4	0.001		

Chickpea and lentils			
Maximum rank	Eigenvalue	Trace statistics	5% critical value
0	—	194.62	47.21
1	0.191	76.91	29.68
2	0.117	7.81*	15.41
3	0.012	0.77	3.76
4	0.001		

Pigeon pea and green gram			
Maximum rank	Eigenvalue	Trace statistics	5% critical value
0	—	99.01	47.21
1	0.109	35.23	29.68
2	0.045	9.84*	15.41
3	0.016	0.09	3.76
4	0.001		

Pigeon pea and lentils			
Maximum rank	Eigenvalue	Trace statistics	5% critical value
0	—	96.77	47.21
1	0.109	32.62	29.68
2	0.039	10.40*	15.41
3	0.012	3.39	3.76
4	0.006		

Pigeon pea and black matpe			
Maximum rank	Eigenvalue	Trace statistics	5% critical value
0	—	270.82	47.21
1	0.258	105.51	29.68
2	0.158	9.95*	15.41
3	0.015	1.05	3.76
4	0.001		

Source: Authors' estimation.

Note: — = data not available. * = statistically significant.

TABLE 6A.3 Price movement for major pulses in prime domestic markets (rupees per quintal)

Major pulses	Wholesale market (mandis)	December 15	November 15	December 14	Percentage of change over previous month (%)	Percentage of change over previous year (%)
Chickpeas	Delhi	4,986	5,229	2,914	−4.65	71.11
	Indore	4,788	4,991	3,064	−4.07	56.27
	Bikaner	4,928	5,025	3,077	−1.93	60.16
Lentil	Kanpur	5,965	6,616	5,818	−9.84	2.53
	Delhi	6,844	6,633	5,806	3.18	17.88
	Indore	6,275	6,647	5,894	−5.60	6.46
Pigeon pea	Gulberga	10,567	11,623	5,174	−9.09	104.23
	Kanpur	8,855	8,780	4,971	0.85	78.13
	Amravati	9,710	9,767	5,106	−0.58	90.17
	Vijaywada	10,343	10,659	4,844	−2.96	113.52
Black matpe	Jalgaon	9,804	10,000	5,375	−1.96	82.40
	Jaipur	10,129	9,400	5,754	7.76	76.03
	Delhi	n.a.	n.a.	6,652	n.a.	n.a.
Green gram	Vijaywada	n.a.	n.a.	7,817	n.a.	n.a.
	Indore	7,738	8,171	7,846	−5.30	−1.38
	Jaipur	7,431	7,679	7,826	−3.23	−5.05

Source: Agriwatch.data (various years).
Note: n.a. = not applicable.

TABLE 6A.4 Johansen's trace test statistic for determining the number of cointegrating vectors

Statistics		Rank	
Maximum rank	0	1	2
Eigen value	.	0.08	0.00
Trace statistic	47.3221	1.81#	0.00

# denotes that the trace statistic at =1 is the value of selected by the Johansen's multiple trace test procedure.

FIGURE 6A.1 Weekly patterns of imports and Wholesale Price Index (WPI) of pigeon pea, 2002–2012

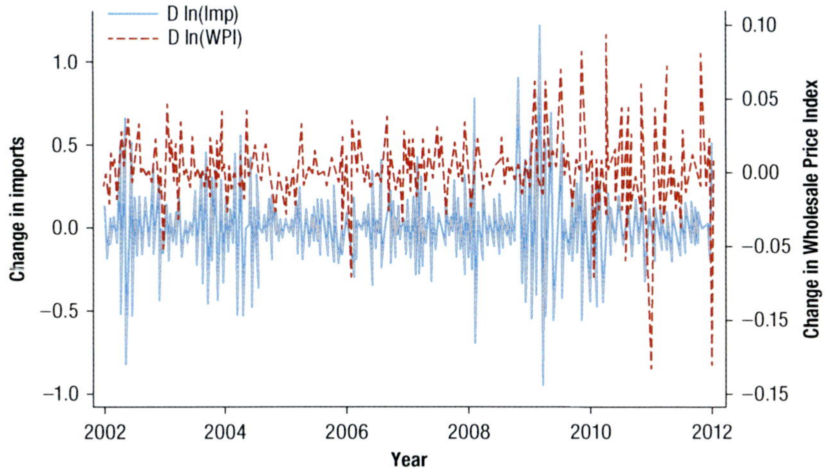

Source: Indian Customs Data and data on wholesale prices from India, Ministry of Finance.

Note: D_ln(Imp) and D_ln(WPI) denote first differences of log (Imports) and log (Prices), respectively.

FIGURE 6A.2 Chickpea imports by India, 2000–2012

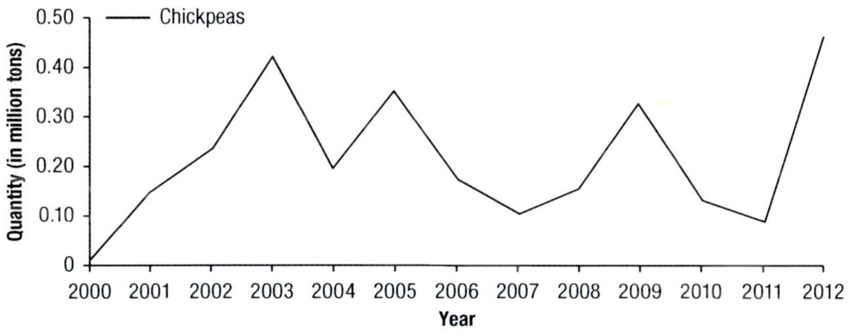

Source: Authors' calculations based on India Customs Data (2000–2012).

FIGURE 6A.3 Pigeon pea imports by India, 2000–2012

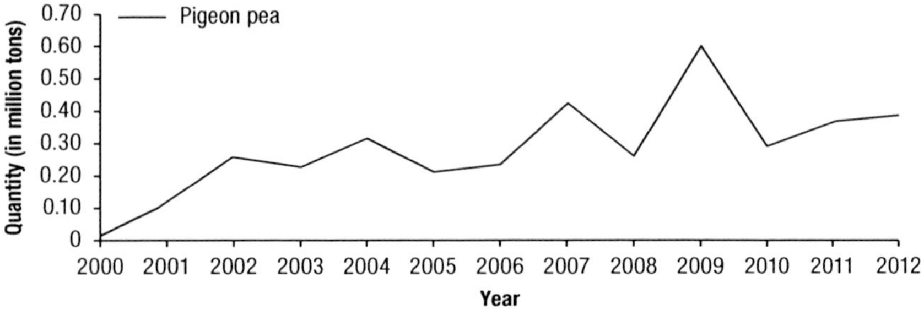

Source: Authors' calculations based on India Customs Data (2000–2012).

FIGURE 6A.4 Imports of black matpe and lentil by India, 2000–2012

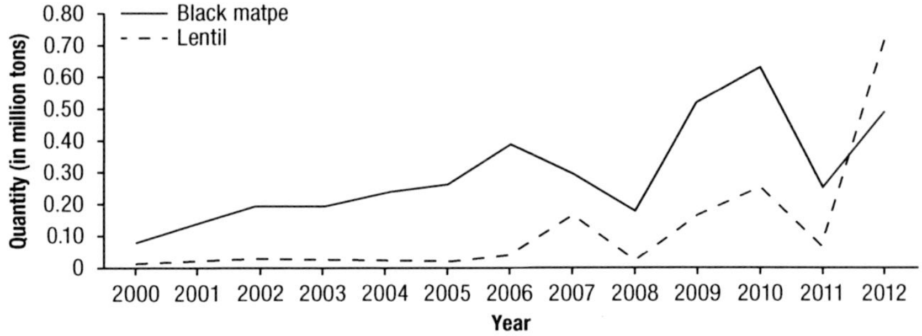

Source: Authors' calculations based on India Customs Data (2000–2012).

FIGURE 6A.5 Imports of green gram and beans by India, 2000–2012

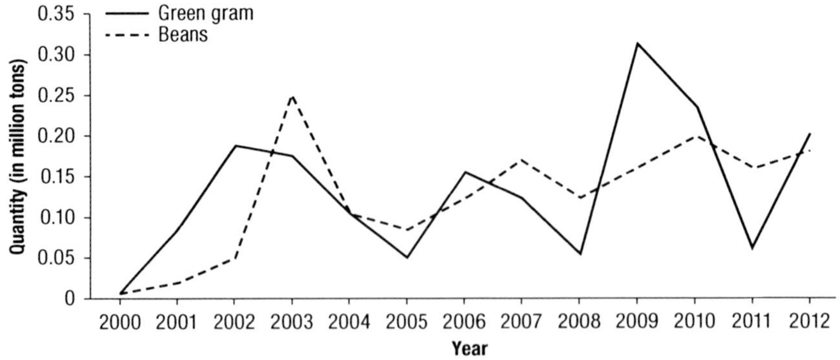

Source: Authors' calculations based on India Customs Data (2000–2012).

Chapter 7

PULSE VALUE CHAIN TRANSFORMATION THROUGH FOOD CONVERGENT INNOVATION FOR A HEALTHY DIET

Laurette Dubé, Srivardhini K. Jha, and John McDermott

India and other low- and middle-income countries (LMICs)—or "emerging" economies as they are also known—have made major investments over past decades into linking agriculture, food, and nutrition to reduce hunger and improve health and other facets of human development (Pingali 2012; Fan and Pandya-Lorch 2012). There have been several good development efforts at pilot scale to support smallholder agriculture and its link to health, nutrition, and community development, as well as efforts to support smallholder agriculture as an input source for food-industry supply chains. Many of these are innovative community-based development efforts that aim to improve livelihoods and have favorable effects on food and nutrition security (see Wiggins and Keats 2013 for pathways and examples). The challenge is to bring some pilot efforts to scale in an economically and socially sustainable way.

Background

Added to this long-recognized problem of scale is a new food and nutrition challenge—namely, managing the dietary transition to avoid obesity and associated noncommunicable diseases. While no country to date has been successful in blocking this trend (Ng et al. 2014), in some countries the trend is worsening rapidly, and India is one of them. While India is experiencing significant economic growth and joining modern markets, it still suffers from a prevalence of nutritional deficiency higher than many other less economically successful neighbors. Home to 40 percent of the world's malnourished children, India is experiencing rising obesity rates, even in its urban slums. India has taken on the dubious reputation of being the "diabetes capital of the world" (Yesudian et al. 2014; Gaur, Keshri, and Joe 2013; Misra et al. 2001). This rapid transition is particularly worrisome for the poor, who fall victim to the triple burden of poor nutrition: they either are hungry or nutritionally deficient, or, if they get enough food that they can afford, it is of poor nutritional quality, high in carbohydrates and, when processed, high in sugar, fat, and salt. As countries like India expand their commercial value chains and markets, this persistent triple burden of nutrition poses a serious challenge to the long-term sustainability of both their human and their economic development outcomes.

We posit that food businesses, situated at the intersection of agriculture, health, and industrial economic systems, can contribute importantly to solving this problem by making nutritional security a cornerstone of both their core business strategies and their corporate social responsibility activities. Food businesses can mainstream nutrition as a driver of commercially successful food innovation, bringing to bear their innovation, communication, and marketing capabilities to modernize traditional food and foster behavioral change, at scale, in the direction of a healthy diet. Food businesses can also bring their entrepreneurial, logistic, and financial capabilities for inclusive and distributed value chains anchored in nutritious agricultural products, small-scale agriculture, and local/regional support for small and medium-size enterprises. By applying their corporate social responsibility, they can partner with government and civil society organizations to support local communities and vulnerable populations with better infrastructure, higher impact, and more resilience.[1] We term such a strategic shift a food "convergent innovation" (CI), and we examine pulses in India as a test bed for laying the foundations of such an approach. Beyond their protein and fiber content, pulses have many health benefits, including blood pressure control, increased satiety, and reduced BMI and risk of obesity, diabetes, and cardiovascular diseases (Boye, Zare, and Pletch 2010).

Notwithstanding these nutritional and health benefits, the food production and consumption patterns in pulses show a worrisome trend worldwide. Consider India, the largest producer and consumer of pulses. As discussed in earlier chapters, pulses have significant nutrition and health benefits. Yet in India, a country with a significant proportion of the population suffering from protein deficiency, their consumption per capita has declined over time, and production and productivity remained relatively stagnant for long periods of time, in part because pulses are less remunerative than grain crops. Consequently, pulse cultivation has been relegated to less productive areas under riskier rainfed conditions, mostly by poor smallholders, potentially entrapping them in a vicious cycle of poverty. Yet, at the same time, India is

1 The analysis here draws heavily from Jha et al. (2014) on convergent innovation.

also the scene of some innovative CI-like initiatives by food businesses that offer promises of solution at scale, including MoPu (More Pulses) and associated programs by the Tata Group.

In addition, India is the site where academic, policy research, and civil society organizations are experimenting to find concrete pathways through which convergent innovation can help integrate the most modern methods of innovation, behavioral and management science, and agriculture-nutrition interventions in hinterland communities for larger and faster twin impact. This chapter reviews key tenets of the CI approach, highlights important features of CI-like initiatives by Indian pioneers, and illustrates further CI strategic deployment opportunities.

Objectives of this Chapter

This chapter has the following objectives: (1) to provide an overview of CI and its ability to bring about behavioral change and value-chain transformation at scale; (2) to describe the pioneering Indian CI initiatives and the Pulse Innovation Platforms; and (3) to discuss the barriers to and enablers for CI and implications they have for policy makers and private enterprises.

Convergent Innovation: An Overview

The concept of convergent innovation has been advanced by several scholars as a solution-oriented paradigm to define new paths of convergence between economic growth and human development that go beyond what has been possible so far in addressing the complex societal problems at the nexus of agriculture, food, and health (Dubé, Pingali, and Webb 2012; Dubé et al. 2014; Jha et al. 2014). CI operates through multilevel and cross-sectoral collaborative platforms to redefine the links between villages and modern value chains. It aims to accomplish this by mainstreaming nutrition and health as drivers of commercially successful food innovation as populations move from subsistence to industrialized modern markets.

This approach is very much in line with similar integrative development strategies suggested by several leading economists to foster broad system-level transformation, including Krugman (1996; 1997); Deaton (2013); Ostrom (2009); and Acemoglu, Robinson, and Woren (2012). This is also consistent with what Reardon et al. (2012) have called "the Quiet Revolution" in developing economies. CI pushes the boundaries of these system-level integrative approaches by simultaneously targeting behavioral change at the individual level and broader societal transformation in both sectoral and intersectoral actions to reach the twin development goals.

As illustrated in Figure 7.1, CI considers the transformation of agricultural commodities through value-addition by the food sector as a key domain where business strategy and investment have a direct impact, both on the economic performance of the sector and on the nutritional quality of the diet (Dubé, Pingali, and Webb 2012; Jha et al. 2014). CI brings to bear the most recent science and practice of understanding individual, organization, and system behavior to create next-generation transformative models of innovation, entrepreneurship, and business engagement. The CI framework relies on a combination of technological innovations in agriculture, farming, food science, nutrition, and health, using social and institutional innovations to scale up and accelerate the transformation of rural and urban communities toward a better convergence of their human and economic development. The approach sees four key enabling conditions: (1) novel scientific insights and methods to inform strategies for incentive design and behavioral change; (2) critical mass in strategic business engagement; (3) community mobilization and cross-sector collaboration; and, finally, (4) the ever increasing digitization of operational and administrative data and digital literacy within and across organizations, value chains, and systems in industrialized societies.

CI nurtures the creation of supply capability and demand conditions for commercially successful food innovation derived from naturally nutritious agricultural commodities in rural and urban contexts, aiming for more distributed value creation and value capture with a bias toward local communities. By addressing both the supply and demand side of the problem, CI aims to reduce the present agriculture-nutrition disconnect and position agriculture as a lasting driver of human and economic development as modernization occurs, achieving a better balance between farm-nonfarm activities in both rural and urban contexts. CI targets the diverse and sometimes conflicting drivers of personal choice and behavior, considering that the individual is at the same time a consumer, a healthcare patient, a family member, and peer, a citizen, and/or a producer. Market or societal investments in agriculture, in food, or in health are only effective if they are consumed by the end user. Therefore, efforts to scale the impact of agriculture and nutrition efforts require deep insight into human behavior. CI convenes actors and institutions from community, education, health, and other relevant social and economic sectors of society to join forces with business to create an enabling and dynamic ecosystem (Figure 7.2). This ecosystem will progressively transform

FIGURE 7.1 Convergent Innovation

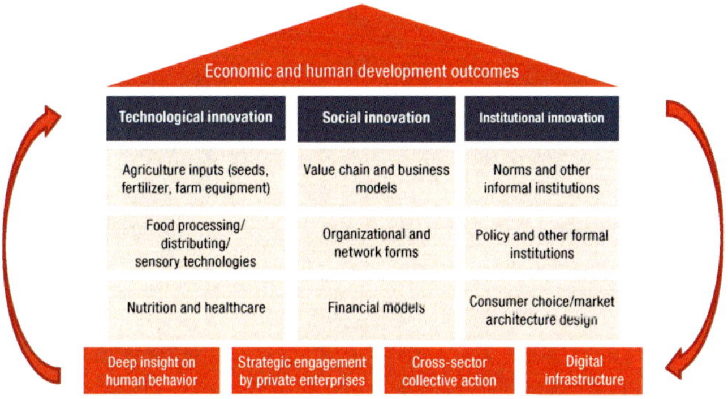

Source: Adapted from Dubé et al. (2014).

FIGURE 7.2 Convergent Innovation as part of commercially successful healthy food systems

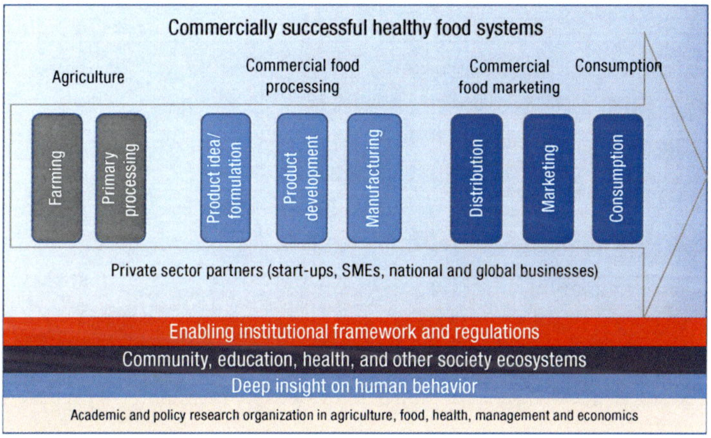

Source: Adapted from Dubé et al. (2014).

institutions such that CI becomes the new standard in a commercially successful and healthy food system.

CI requires the engagement of the private sector in a deeper and more strategic way than current private-public partnerships (IDRC 2012) to bring solutions to scale for food security in all its facets. Food CI brings business capacity for technological innovation, entrepreneurship, and supply chain/ market development to simultaneously enhance the nutritional profile of food products and link them back to agrarian communities to chart a development course that is scalable and resilient. By *food businesses* we mean the full scope of enterprises, from micro to small and medium-size enterprises operating in both formal and informal economies all the way up to national and multinational corporations. In a country like India, the food sector currently comprises more than 36,000 registered food-processing units and almost as many firms, together constituting close to 40 percent of the workforce.[2] In addition, there are many unorganized units with a greater share in employment but much smaller share in output. These, along with agriculture, contribute more than 17 percent of GDP (Gulati et al. 2012). The question is: How can these start-ups and small manufacturing enterprises (SMEs) be set on a course of convergent innovation? Can they serve as a catalyst for modernizing traditional foods, making them functional, aspirational, commercially successful, and better equipped to compete with incoming nutrient-poor, calorie-rich Western foods?

While CI provides a framework for collaborative and transformative projects, its practitioners recognize that each CI project tackles a unique problem that needs to be addressed by a specific set of partners. In the field of CI for food, each project is designed to create private or public value, or both, by assembling a roadmap for change at scale to address societal problems within the context of specific agricultural commodities. For example, CI for pulses could offer a food-based solution to improve the nutritional quality of diet and reduce stunting and diabetes. Success hinges on partners being able to establish a high level of trust, recognize interdependencies, create shared goals, and execute a project that transcends organizational boundaries. In sum, CI is not a simple process but one that is iterative and requires strategic commitment from all partners.

CI Initiatives

The Case of Tata MoPu/i-Shakti

The Tata More Pulse (MoPu) initiative, jointly spearheaded by Tata Chemicals and its agrochemical subsidiary, Rallis India, came out of Tata Sons's long-term strategic initiative on "Where is agriculture headed?" Under this wide umbrella, the company decided to address the pulse value chain, given the nutritional importance of pulses and the inefficiencies that plagued the value chain and made India a net importer of pulses. This move is one of

2 "Food Processing," www.makeinindia.com/sector/food-processing/.

the several business-driven initiatives that are expected to transform value chains and food innovations. Readers should, however, note that the various claims of impact and outreach made in the description of this case study are not yet based on rigorous formal evaluations entailing baselines and appropriate control groups. Rather, they rely on information provided by the promoters and partners in the initiative and by a study performed by the Federation of Indian Chambers of Commerce and Industry (FICCI), an association of business organizations.

Public-Private Partnership Arrangement

The MoPu initiative was piloted in the Pudukottai district of Tamil Nadu as a public-private partnership with the government of Tamil Nadu and sought to streamline every stage of the pulse value chain—from production to processing to marketing. Tata Rallis created a team of extension workers to provide timely access to, and diffuse innovations in the use of, high-quality pulse seeds, fertilizers, and pesticides to the farmers. The extension workers interacted closely with the farming community, advising them on scientific farming techniques such as soil testing, irrigation methods, and water harvesting. To effectively reach all the farmers, Rallis India leveraged a technology platform by Tata Consultancy Services (TCS), called mKRISHI. The platform is aimed at allowing extension workers to resolve farmers' queries within 24 hours using the backend support of agri-experts.

After the feasibility studies the program was rolled out in the states of Maharashtra and Karnataka. The government of Maharashtra signed a memorandum of understanding (MoU) with Rallis India to roll out MoPu as a public-private partnership. Under this arrangement, the package of best practices offered by Rallis India was subsidized by the government, with the goal of boosting pulse production in the state. Furthermore, an amendment in the Agricultural Produce Marketing Committee (APMC) Act by the government of Maharashtra in 2006 facilitated the private markets to actively participate in direct marketing. This enabled farmers to sell their produce in the open markets and not be constrained by the APMC market yards. With this institutional support, the project expanded to more than 35,000 farmers in 372 villages across three districts of Maharashtra (FICCI 2013). Rallis India also provided a buyback option by setting up a procurement center that was expected to offer more competitive prices than the local market.[3]

---

3 According to FICCI (2013, 37), some farmers stated that at the Rallis procurement center the rates given for the produce are at par with the prevailing rates at the nearby *mandis*.

Farm-Level Engagement

Based on an FICCI (2013) study, farmers across three districts reported a 30–65 percent increase in productivity and a similar increase in their income. They also reported a 20–30 percent savings on fertilizer expenditure due to efficient soil testing. Note that these figures are not based on a rigorous evaluation study that accounts for treatment and control and/or before and after comparisons. The scale of the project helped forge partnerships with banks, such as Housing Development Finance Corporation (HDFC) and State Bank of India, which opened zero-balance accounts for the farmers. This facilitated the direct transfer of money from Rallis to farmers for the produce procured and is claimed to have reduced delays, ensuring benefits to the farmers. Although more formal process and outcome evaluations have yet to be performed, with baseline assessment and appropriate control groups, early results do indicate the potentially main drivers of outcomes that appear to be the timely provision of adequate agricultural information on appropriate seeds and fertilizers, which translate into higher agricultural yields, as well as reduced input costs tied to better quantity control and/or lower price per unit. Note that subsidized inputs may have driven the results at least in part.

Food-Processing Engagement

Turning to the food side of this CI project, the Tata Chemicals market procured pulses under the i-Shakti brand, transforming a loosely sold product category to a branded one. Furthermore, i-Shakti pulses are unpolished. According to Tata Chemicals, being unpolished means that while processing i-Shakti pulses, no marble powder (which is harmful to the intestines), oil polish (which adds fat), water polish (often using unknown sources of water), or leather belt polish (which uses animal skin) is added or used.[4] This ensures that the pulses retain a higher protein content, making them a healthier product for consumers at similar price points as polished pulses, a claim made by Tata Chemicals.[5] Industry observers claim that the undertaking of CI-like projects by the Tata Group has affected many in the chain. For Rallis India, it has opened up new markets for their seed, fertilizer, and pesticide technologies. For Tata Chemicals it has provided a source of marketable high-quality pulses. For farmers it has potentially delivered productivity gains, savings on inputs, commissions, and weighing charges (FICCI 2013), leading to

4 Tata Chemicals, www.tatachemicals.com/products/pulses.htm.

5 The i-Shakti pulses are sold at a marginal premium of about 10 percent more than the loose variety.

increased income. For the respective state governments, the initiative could boost pulse production and net regional agricultural output. Finally, the consumers have benefited from access to branded and potentially fewer contaminated and healthier pulses.

However, it should be noted that close to half of the costs of the initiative were covered by the state government, which is a fiscal challenge if programs like this are to be scaled up significantly. It is also noteworthy that the FICCI (2013, 40) study mentions a few areas in this initiative that require attention. These include (1) farmers' complaints about lack of inclusiveness, whereby only a few selected farmers in a village get free inputs and subsidies; (2) cases of untimely supply of project input that is at times not synchronized with the cultivation schedule; and (3) farmers' complaints about quality seeds not being part of the kit of inputs distributed as part of the initiative.

What does the MoPu initiative indicate in terms of a CI? The initiative does involve strategic engagement by large private enterprises, in this case, Rallis India and Tata Chemicals, which could be crucial to bringing in formal market mechanisms and building capacity in the informal sector. It also mobilizes the farming community and requires the collaboration of actors from multiple sectors: government entities and private firms in the agricultural and financial sectors. Finally, the initiative leverages the digital platform, mKRISHI, to scale the effort. In other words, it has three key enablers of CI in place that were discussed earlier.

Using these key enablers as a platform, the MoPu initiative includes a bundle of accompanying innovations. Rallis's technological innovations in seeds, fertilizers, and pesticides are supported by the social process innovation that creates a community of pulse farmers equipped with scientific farming techniques. The institutional innovation, in the form of an amendment to the APMC Act and the subsidy for the package of practices, creates an environment that can facilitate pulse production and marketing. The financial innovation of zero-balance accounts among millions of unbanked farmers allows cashless transactions, reducing transactional leakage.[6] Finally, the collaboration between Rallis and Tata Chemicals is an organizational innovation that integrates the rural community into the industrial value chain. In the times ahead, as the program will be in place for some time, evaluation studies will be able to assess the true impact of the program.

---

6 The government of India has launched a large-scale program of financial inclusion where several zero-balance accounts have been set up. The program is called Jan Dhan Yojana.

Odisha Pilot

A new CI Odisha pilot has been introduced as a collaborative project across four organizations: PRADAN, an NGO with extensive reach to agrarian communities; iKure, a social business that aims to provide affordable and quality primary healthcare services in rural areas using modern technology; McGill Centre for the Convergence of Health and Economics (MCCHE), a research center dealing with CI; and the International Food Policy Research Institute (IFPRI). As part of the CI, because Odisha has potential for cultivation of underused pulses (Joshi, Birthal, and Bourai 2002), lentil and chickpea will be promoted among farmers with suitable high-yielding varieties and will be disseminated through PRADAN's network of community service providers (CSPs) and self-help groups (SHGs). Second, behavioral change will be stimulated by generating awareness about nutrition, especially maternal and child nutrition. Third, a set of context-specific digital content will be created, in partnership with IFPRI and other nutrition domain experts, and embedded into the community through PRADAN's field network of CSPs and SHGs. Finally, iKure will provide exposure to preventive and curative healthcare through health camps, remote consultation, and technology-backed monitoring. iKure's health intervention will provide relevant preventive healthcare measures and access to qualified medical professionals for timely diagnosis and appropriate disease treatment.

India Pulse Innovation Partnership

The Pulse Innovation Partnership (PIP) is a global alliance of public and private organizations, civil society, and academia, committed to increasing the consumption of pulses in the developing and developed world by creating novel pulse-based processed foods. It is spearheaded by MCCHE, CGIAR, the Global Pulse Council (GPC), and Pulse Canada. PIP hinges on open innovation (Chesbrough 2003) and "co-opetition" (cooperative competition) (Brandenburger and Nalebuff 2011), bringing together actors from all stages of the pulse value chain to work in a collaborative model. As a CI effort, its purpose is to bring pulse-based food products at scale to the market. The project involves bringing together food-processing companies (primary, secondary, and tertiary), trading companies, and marketing companies in the pulse value chain to develop innovative pulse-based products.

Through a flexible partnership model that incentivizes and supports pulse-based food innovation, PIP plans to provide three types of knowledge services to food companies: innovation services, marketing services, and policy support. In the case of pulse-based products, PIP will focus on modernizing them to better compete with nutrient-poor, calorie-rich foods. The food innovation and entrepreneurship capacity of PIP are to be further linked with efforts in promoting pulse production, community mobilization, population health, and health systems improvement to target behavioral change and ecosystem transformation at scale in rural and urban communities, value chains, and markets. The aim is that through the participation of several small, medium-size, and large Indian food companies, these could eventually lead to many traditional pulse-based foods and grassroots food innovations flowing into industrial innovation pipelines, opening up a new category of processed foods. The initiative promises to expand the choice of pulse-based products available for consumption, potentially boosting demand and providing a fillip to production.

Conclusion

This chapter has offered an overview of convergent innovation (CI) and outlined three different projects at different stages of development, each convergent in nature and aimed at transforming a particular segment of the pulse value chain. Each of these projects is a small step toward the same overarching goal. Each project involves a relatively small number of partners and focuses on improving a subset of the overall ecosystem. This modular approach keeps CI complexity at manageable levels and allows validation of the CI concept itself. When woven together these projects will likely provide a roadmap for ecosystem transformation. However, as the chapter discusses, the projects are complementary and together they should contribute to the goal of transforming the pulse value chain.

Two of the initiatives are still in the pipeline. The India Pulse Innovation Partnership, which provides a platform for developing innovative pulse-based food products, will engage with food companies of all sizes, including small firms in rural areas and large vertically integrated companies with expansive supply and distribution networks. These food innovators could become connected to the country's emerging pulse-producing hubs, such as the smallholder farming communities developed through the other two CI efforts, the established MoPu initiative and the soon-to-launch Odisha pilot. Such links will not only likely provide an impetus to the rural economy and embed the smallholder farmers into industrial value chains, but will also potentially allow food-processing companies to work with the farming community to source the right variety and quality of pulses. Linking these and other projects in the pulse domain will likely lead to at least three important outcomes:

- Economic growth for smallholder farmers through their integration into mainstream economic activity,
- Decentralization of growth from urban to rural areas through the proliferation of a number of small and medium enterprises, and
- Improvement in population health status from behavioral change cues and an array of healthy food choices in the form of pulse-based products.

Creating these links and scaling them up will entail a new level of complexity and require the active involvement of government departments and a supportive policy framework. First, direct procurement from farmers might be needed. It is worth noting that the amendment to the APMC Act in Maharashtra has played a crucial role in the MoPu initiative. Second, if agriculture, nutrition, and health challenges are to be addressed together, multiple departments in the government need to work together, overcoming their current isolated approach. Third, incentives need to be put in place for private enterprises to engage in nutrition-sensitive food innovations—for example, including R&D tax credits or challenge grants. And finally, there is a need to create incentives for consumers to procure nutritious products.

In sum, there is a need for institutional innovations that better aligns national and state-level governmental policy and programs in agrifood and social sectors in their ability to harness the private sector's ability to scale up supply and demand for nutritious food for all, involving both local and regional SMEs as well as large national and multinational corporations. The exact form of these large-scale transformative policy interventions has yet to be defined, and this will present challenges and opportunities for policy makers, businesses, and actors throughout society for years to come.

References

Acemoglu, D., J. A. Robinson, and D. Woren. 2012. *Why Nations Fail: The Origins of Power, Prosperity, and Poverty*. Vol. 4. New York: Crown Business.

Boye, J., F. Zare, and A. Pletch. 2010. "Pulse Proteins: Processing, Characterization, Functional Properties, and Applications." *Food and Feed* 43 (2): 414–431.

Brandenburger, A. M., and B. J. Nalebuff. 2011. *Co-opetition*. New York: Random House.

Chesbrough, H. W. 2003. "The Era of Open Innovation." *MIT Sloan Management Review* 44: 35–41.

Deaton, A. 2013. *The Great Escape: Health, Wealth, and the Origins of Inequality*. Princeton, NJ, US: Princeton University Press.

Dubé, L., S. K. Jha, A. Faber, J. Struben, T. London, A. Mohapatra, N. Drager, C. Lannon, P. K. Joshi, and J. McDermott. 2014. "Convergent Innovation for Sustainable Economic Growth and Affordable Universal Healthcare: Innovating the Way We Innovate." *Annals of the New York Academy of Sciences* 1331 (1): 119–141.

Dubé, L., P. Pingali, and P. Webb. 2012. "Paths of Convergence for Agriculture, Health, and Wealth." *Proceedings of the National Academy of Sciences* 109: 12294–12301.

Fan, S., and R. Pandya-Lorch. 2012. *Reshaping Agriculture for Nutrition and Health*. Washington, DC: International Food Policy Research Institute (IFPRI).

FICCI (Federation of Indian Chamber of Commerce and Industry). 2013. *Improving the Productivity of Pulses through Public Private Partnership*. New Delhi.

Gaur, K., K. Keshri , and W. Joe. 2013. "Does Living in Slums or Non-Slums Influence Women's Nutritional Status? Evidence from Indian Mega-Cities." *Social Science Medicine* 77: 137–146.

Gulati, A., A. Ganesh-Kumar, G. Shreedhar, and T. Nandakumar. 2012. "Agriculture and Malnutrition in India." *Economic and Political Weekly* 33: 74–86.

IDRC (International Development Research Centre). 2012. *Implementing Public-Private Partnerships in Agriculture: Creating Value for Smallholder Farmers*. Ottawa, Canada: IDRC.

Jha, S. K., J. McDermott, G. Bacon, C. Lannon, P. K. Joshi, and L. Dubé. 2014. "Convergent Innovation for Affordable Nutrition, Health, and Healthcare: The Global Pulse Roadmap." *Annals of the New York Academy of Sciences* 1331 (1): 142–156.

Joshi, P. K., P. S. Birthal, and V. A. Bourai. 2002. *Socioeconomic Constraints and Opportunities in Rain-fed Rabi Cropping in Rice Fallow Areas of India*. Patancheru, India: International Crops Research Institute for Semi-Arid Tropics (ICRISAT).

Krugman, P. R. 1996. *The Self Organizing Economy*. Blackwell Publishers.

———. 1997. *The Age of Diminished Expectations: U.S. Economic Policy in the 1990s*. Third edition. Cambridge, MA, US: MIT Press.

Misra, A. R., M. Pandey, J. R. Devi, R. Sharma, N. K. Vikram, and N. Khanna. 2001. "High Prevalence of Diabetes, Obesity, and Dyslipidaemia in Urban Slum Population in Northern India." *International Journal of Obesity Related Metabolic Disorders* 25: 1722–1729.

Ng, M., T. Fleming, M. Robinson, B. Thomson, N. Graetz, C. Margono, E. C. Mullany et al. 2014. "Global, Regional, and National Prevalence of Overweight and Obesity in Children and Adults during 1980–2013: A Systematic Analysis for the Global Burden of Disease Study." *The Lancet* 384 (9945, August): 766–781.

Ostrom, E., 2009. "Analyzing Collective Action." Keynote papers presented in the Plenary Sessions of the 27th Conference of the International Association of Agricultural Economists. August 16–22, 2009, Beijing, China.

Pingali, P. L. 2012. "Green Revolution: Impacts, Limits, and the Path Ahead." *Proceedings of the National Academy of Sciences* 109 (31): 12302–12308.

Reardon, T., K. Chen, B. Minten, and L. Adriano. 2012. *The Quiet Revolution in Staple Food Value Chains: Enter the Dragon, the Elephant, and the Tiger.* Mandaluyong City, Philippines: Asian Development Bank; Washington, DC: International Food Policy Research Institute.

Wiggins, S., and S. Keats. 2013. "Smallholder Agriculture's Contribution to Better Nutrition." Overseas Development Institute. Report Commissioned by the Hunger Alliance. London, UK. March. Accessed September 2016. https://www.odi.org/sites/odi.org.uk/files/odi-assets/publications-opinion-files/8376.pdf.

Yesudian, C. A. K., M. Grepstad, E. Visintin, and A. Ferrario. 2014. "The Economic Burden of Diabetes in India: A Review of the Literature." *Globalization and Health* 10 (80): 1–18.

Chapter 8

CONCLUSIONS AND WAY FORWARD

Devesh Roy, P. K. Joshi, and Raj Chandra

The still-unresolved problem of inadequate access to good sources of nutrition has significant implications for human capital formation and overall economic growth. The relationship between poverty and food and nutritional insecurity defines a vicious circle whereby poor people cannot afford a sufficiently nutritious diet. Consequently, they and their children often suffer from various physical handicaps that impair their ability to be fully productive in ways that might alleviate their poverty. In India this relationship has significant implications, as the country has a relatively high rate of poverty and indeed the largest number of malnourished people in the world. Because pulses not only have beneficial nutritional properties but can also enhance agricultural productivity through improved technology (thereby improving the incomes of poor smallholders and reducing the price of the product to poor consumers), they have a significant role to play in agriculture for nutrition and health.

During the past half century, India has come a long way in enhancing food availability. The adoption of new technologies during the Green Revolution brought about a significant change in the production of food grains. Ironically, while India emerged as a grain-surplus state, the studies in this book show that a side-effect of that success has been a steady decline in other nutritionally important crops, including pulses. In particular, the persistence of the demand-supply gap in pulses despite significant budgetary allocations by the government on programs for technology development, improvement, and diffusion (see Chapters 3 and 4) implies that improving the performance of pulses remains an important policy issue in India.

As discussed in Chapters 1 and 2, although the main role of pulses in the Indian diet as a provider of protein has moderated, pulses continue to be quite important in relation to other noncereal protein sources. Moreover, compared with those other sources, pulses remain cheap. Given the protein insufficiency in the diets of far too many Indians, policies discussed in this book that could reverse the declining role of pulses in protein intake have important implications for dietary health. A number of analytical findings and policy implications for the pulse sector, in relation to other crops such as cereals, emerge from this edited volume. In Chapter 2's detailed presentation of pulses' consumption dynamics, the long-term decline in per capita consumption of pulses is documented, but it is also made clear that there has been a rising aggregate demand for pulses. That demand owes to several factors, such as increases in population, in urbanization, and in the demand for processed products that use pulses as an ingredient. Despite this rising demand, as Chapter 3 highlights, pulses have suffered marginalization as a crop, pushed out of traditional growing areas into nonirrigated, rainfed areas. This movement has had significant implications for the supply of pulses as well as for technology development (discussed in Chapters 3 and 4).

The dynamics of pulse processing have been dominated over the past decade by the fast growth in the number of mills and the magnitude of output of the organized sector, accounting now for three-quarters of marketed processed pulses. The organized sector operates comparatively large plants, although the technology used is often not as advanced as that of international producers. The unorganized sector has a large number of small-scale mills using grossly outdated technology. The potential of both pulse growth and the processing sector has been suppressed by the fact that the sector tends to have few direct links with farmers owing to several reasons, such as marketing regulations, as well as the relatively small volume and inconsistent supply of raw materials from small-scale traders that limit the scope for processors to develop, expand, and operate at full capacity.

Chapter 5 shows that processing costs and processors' margins are high, accounting for about two-thirds of the difference between farmgate prices and consumer prices. Therefore the promotion of processing-cost-cutting policies (better processing technology, organizational arrangements to link producer organizations to processors) have the potential to bring about a lower consumer price. Lack of processing has resulted in fewer differentiated products. Product differentiation in fact is one factor with the potential to increase consumer demand for pulses and pulse-based products and thus a way to improve the returns to farmers and thereby improve the supply of pulses as well.

Despite India's remaining the largest producer and consumer of pulses, its domestic production now consistently falls short of demand. Given the demand-supply gap, imports of pulses have been increasingly significant. Based on the demand-supply projections for pulses to 2030 presented in Chapter 2, without substantial changes in domestic pulse farming, India may need to import a minimum of 3 million to 4 million tons of pulses every year in times to come. With reliance on imports rising, Chapter 6 captures the dynamics of the pulse trade and finds that although imports have expanded significantly, they still have not been large enough to stem domestic price increases significantly. Part of the reason for this limited role of the sizable trade has been the margins along which trade has moved over time. As Chapter 6 shows, until recently trade has expanded more on the intensive margin, with only a few exporting countries dominating many pulse varieties, and only more recently has it expanded on the extensive margin as India has become a more important market destination. This is reflected in the recent entry of several East African countries into the Indian pulse market.

India should specialize in those pulses where it has a comparative advantage. In pulses, the reliance on trade is nuanced because the demand from India is large and the set of possible exporters of pulses to India, globally, is still small. In effect, while relying on imports, there might not be many options (in terms of the number of countries) to buy from in the world market. Although it is not our intent to favor self-sufficiency, for some pulses, domestic production might remain the primary available source of supply, especially in the short run. Chapter 7 looks at the development of the pulse sector from a convergent innovation (CI) perspective. The CI examples highlight the potential for such innovations to improve outcomes for the pulses sector. For a CI approach, exploiting the market potential of pulses is quite important, and in this regard, the pulse processing sector can play an important role. Innovative secondary processing could bring a wide variety of new products into the markets with genuine product differentiation. Many value-added pulse-based products are already in the market, but unfortunately, the necessary consumption data to measure the demand for them is not currently available. Future research on pulse consumption should focus on these value-added products.

In the remainder of this chapter, we summarize the messages that emerge from the preceding studies on a variety of issues related to India's pulse sector. We then highlight the way forward for pulses in India—that is, ways to enhance the gains made in pulses in recent times through further changes in policy and other areas. Broadly, the three priorities related to the way forward concern (1) improving technology for the producers in increasingly marginalized areas; (2) adjusting the trade in pulses with the needs of poor consumers in mind, to the extent possible, given that there are only a few pulse exporters; and (3) developing the value chain to better link consumers with producers, including development of the pulse processing sector.

Key Messages from the Assembled Studies

Reorientation of Pulses Production Systems

The government of India has tried through various policy measures to recover the ground lost in pulse production after the Green Revolution, which resulted in the dominance of the cereals sector. There is clear evidence that pulses have been pushed into marginal environments and into nontraditional areas, and policies and institutions need to respond to these dynamics. Realistic options for improving the outcomes for the pulses sector need to take into account the centrality of the cereals sector in the production systems. For example, pulse growing technologies should provide for the option of using rice fallow systems, which means adjusting to the need for short-duration and very-short-duration pulse crops. In addition, pulses are needed for planting as intercrops along with certain noncereal crops such as soybean—for example, in parts of Maharashtra. Technology development, production systems, and extension services must all respond to these needs of the pulse sector.

There remains a need to bring additional area under pulse cultivation. As detailed in Chapter 3, the total acreage under pulses in India has remained unchanged over a long period of time. Even though pulses have performed well since 2015, with record outputs, the need to bring additional area under pulses is reflected in the July 2016 policy announcement of the government to lease land in Myanmar and in African countries (for example, Mozambique) to grow pulses for export to India. The regions of the country where pulses have moved tend to lack irrigation, making pulses an overwhelmingly rainfed and hence risky crop. Several suggestions from the book's analysis follow for the supply side of pulses. Avenues for area expansion under pulses need to be explored and to be realized, including one immediate candidate: expansion of pulses by using fallow lands.

Utilization of rice fallow lands. In a land-scarce country like India, how can extra land for pulses become available? The analysis in Chapter 3 shows that the norm has been for pulses to be displaced from land once it has been improved through access to irrigation. The discussion in Chapter 4 elaborates on the use of rice fallow lands to increase pulses production significantly. Rice fallow systems are quite widespread and can be suitably used in many areas. Several improvements in technology, extension, and markets need to be in place before the use of rice fallow lands for pulses can happen on a large scale. Pande, Sharma, and Ghosh (2012), who assessed the performance of chickpea in the rainfed rice fallow land of Madhya Pradesh and Chhattisgarh, found mixed responses. They found that the farmers experienced major constraints such as biotic and abiotic stresses and that farmers exhibited poor crop management practices and lacked awareness about modern methods, including quality seeds. Hence, it follows that the introduction of pulses into rice fallows should be accompanied with sufficient extension services (NAAS 2013).

As presented in Chapter 4, there already are varieties of pulses developed for rice fallow systems. Research at ICRISAT, for example, looks at the suitability of the developed varieties at the state level from the perspective of rice fallows utilization. With the technology already in place, there is a greater role for extension for expanding the acreage under these varieties. Note that short-maturity rice varieties may entail some yield reduction compared to longer maturity rice, which may detract from the incentives for their adoption in the rice-pulse cultivation cycle, particularly if farmgate prices of pulses remain unattractive.

Taking pulses to new areas with watershed development. Notwithstanding the concentration of pulses in the rainfed areas, in order to reduce intertemporal variability in yields and also make pulses a more attractive crop, pulses growing in nonirrigated areas should be complemented with an uptake in the irrigated areas. In districts where watershed development has succeeded in recent years, attempts should be made to promote pulses cultivation. Admittedly, the evidence in this book (Chapter 3) shows that access to irrigation leads to a movement away from pulses, so expanding pulse production in areas with watershed development may be possible only if consistently high yields are observed and the relative price distortions that favor cereals and some other crops are removed. Since pulse prices have been consistently high but pulse adoption has not taken off in irrigated areas, it seems that price transmission to the farmer level has been quite limited. The recent episodes of high prices in pulses and record production since 2015 indeed provide an excellent opportunity to foster greater production of pulses in the non-rainfed areas, particularly those where watershed development has most recently taken place.

Technology Interventions

Chapter 4 points out that although technological progress has been made in pulses, it has not been large enough to change the outcomes in the pulses sector to a substantive extent. Sufficient scope remains for taking pulse varieties to the next level by using more of the genetic base and customizing technologies to match the realities of the pulse sector in the cropping complex. Singh and Saxena (2016), while analyzing the yields of different types of pulses in India, show that yield gaps are very high, ranging from 75 percent in lentil to

224 percent in green gram. They suggest that the underlying reasons for these gaps are mainly poor seed quality and poor management practices.

Successes in raising pulse productivity to date seem to be too few and concentrated in just a few pulse varieties—for example, chickpeas—in spite of the development of several varieties by India's public-sector research system. In fact, there have been stumbling blocks in the pulse sector not only in technology development but also in uptake postdevelopment. Hybrid pigeon pea, for example, which was discussed in Chapter 4, shows the limits of a promising technology that has not been adopted extensively owing to, among other reasons, a lack of extension services (Niranjan et al. 1998). The nationally representative situation assessment survey, conducted in 2003 and again in 2013 by National Sample Survey Organization (NSSO), has identified a continuous decline in the use of public extension services (Birthal et al. 2015).

Moreover, in the case of pulses, there is a stark lack of private-sector participation. In both technology development and in extension, the private sector is conspicuous by its absence. For crops where the private sector has become important, such as maize and pearl millet, the diminishing role of public extension services has been compensated for by increased private extension (Feder, Birner, and Anderson 2011). The supply-push policies for research do not seem to have been lucrative for the private sector. Although it does not follow directly from the analysis in this book, it is possible that a demand-pulled research policy, under which with a level playing field the private and public sectors would get rewarded if they developed technology based on predetermined traits, would make a helpful difference.

The case of wheat in India testifies to the potential for success of public sector–driven research, development, and dissemination. The case for public-sector research on and dissemination of improved pulses technology is even stronger for several reasons. First, technology development in pulses involves social benefits that exceed the private benefits driven by profit. Among protein-rich foods, pulses have the lowest carbon and water footprint. In addition, as discussed in Chapter 1, pulses improve soil health by naturally fixing atmospheric nitrogen; growing pulses reduces the need for application of nitrogenous fertilizer, especially urea, in the subsequent crop. Technology development in pulses would improve soil fertility and provide valuable environmental services. Lower usage of fertilizer, pesticide, and irrigation make pulses an environmentally sustainable crop group. Saddled with a huge fertilizer subsidy burden and food safety issues from excessive chemical use in farming, India can benefit greatly from these roles of pulses. Technology development in pulses would have social gains that far exceed private gains.

Consequently, the markets in pulse R&D can fail and public-sector research can be advocated. The public-sector research and development, however, has been in decline for a long time, and the superior quality research undertaken by the private sector is limited to crops that are commercially more attractive, such as maize, cotton, and vegetables. Second, because pulses have been pushed to marginal environments, willingness to pay for new technology is often limited. Also, the flexibilities in selecting the traits in a technology are driven to a large extent by the relationship with competing crops.

India has a comparatively well-developed private sector in seed development, at least in comparison with the average developing country, which caters even to some minor crops like pearl millet. In this context the near complete absence of the private-sector seed producers and extension services in pulses development is striking. To some extent, the reason for this low engagement may be structural: with the exception of pigeon pea, pulses do not outcross, hence the scope for private-sector investment in hybrids and in seed production may be limited. With the absence of hybrids and low seed replacement rate, appropriating the returns on R&D investment is difficult. Examples from cotton, maize, and pearl millet in India show that the dissemination of technology is wider and faster for private-sector seeds because the profit motive creates incentives for private extension. In order to maximize adoption, the private-seed suppliers provide extension services themselves, often using the input suppliers as the agents on the ground (Asare-Marfo et al. 2010). Thanks in part to this practice, improved maize, pearl millet, and vegetables in India have all experienced rapid uptake by farmers once they have become available. Unfortunately, this has not been the case for pulses. An important policy question, therefore, is how to enhance the engagement of the private sector in the development of pulses technology and its dissemination.

Pricing Policies

As discussed in Chapter 3, as a result of relative price distortions, with cereals receiving support prices backed by government procurement, pulse farmers are subject to disincentives. If left uncorrected, these price disincentives could reverse some of the gains made in pulse production in recent years, although, as discussed in Chapter 4, there are a number of nonprice factors that adversely affect the supply responsiveness of pulse producers. Chapter 3 highlighted the role of limited or no procurement of pulses, but argued that the implementability and effectiveness of a large procurement program in pulses cannot be taken for granted and suggested the need for further research before a policy stance on this matter becomes clear. The caveat follows from the experiences in historically procured cereal crops where limited coverage (6 percent) of farmers has been achieved and the observation that the procurement program tended to benefit large farmers.

In addition, the role of Minimum Support Price (MSP) as a focal point aiding tacit collusion among traders, thereby penalizing farmers, was discussed. It was noted that when the government procurement price is lower than the market price, it actually works as a tax on the farmers (Vyas 2003). It was pointed out that to be effective, the procurement price would need to be quite high, raising concerns about the fiscal feasibility and sustainability of a large procurement program. Overall, the stance of the book is that although there is a theoretical economic rationale for introducing parity with cereals as a second-best policy, there are significant practical challenges and nonprice factors hindering the supply response of pulses producers that need to be assessed before formulating a policy on price support and procurement.

On the side of consumer pricing, the analysis presented in Chapter 2 relying on Chakraborty, Kishore, and Roy (2016) for assessing the utility of consumer subsidy shows that the impact of the subsidy on household pulse consumption is of small order. The effective changes were not large enough to bring about any sizable difference in pulse consumption or, by extension, in protein intake. Moreover, the small impact of these subsidies in the face of falling demand due to shifting preferences is uniformly the case across different types of pulses.

Going forward, several different tasks regarding pricing in pulses remain to be tackled. Apart from studying the issue of support prices and procurement, steps need to be taken to ensure better transmission of consumer prices to the prices producers can receive. This has been an acute problem in recent episodes of pulse price spikes as discussed in Chapters 3 and 5, as producer prices continued to be benchmarked to the MSP while the retail prices increased many times. Fixing this problem would require establishing direct farm-to-fork or firm-farm links. The role of farmer producer organizations, which balances the bargaining power toward farmers, can be instrumental in this regard (see Chapter 5), as is the need to reform regulations that constrain direct marketing by farmers. The role of processing too, with its backward links, is very important in this regard. The promotion of improved processing technology could be important in reducing processing costs, which amount to a large share of the final consumer price. In addition, the discussion of convergent innovations in Chapter 7 suggests that private-public cooperation has a potential for transforming the pulses sector.

Addressing Consumption Issues

The discussion in Chapter 2 describes the unique situation in which the per capita consumption of pulses has been declining continuously while aggregate demand has been simultaneously increasing due to rapid population growth. Moreover, as shown in Chapter 2, over time the reduction in per capita consumption of pulses has been uniform across all household categories (wealth categories) and across all regions of the country. Going forward, policies that can affect the ultimate market price to consumers are likely to be quite important. It also showed significantly rising consumption of processed items in the food basket. Some of the processed food items include pulses as well. Note that the consumption surveys in India do not account for secondary processed pulses that could be accounting for the significant increase in demand for pulses. In other words, the consumption of processed pulses is underestimated in the NSS and other consumption surveys.

With the shift in consumption toward high-value items and processed food, pulses processing assumes an important role. If the objective were to increase pulse consumption, avenues for greater availability of processed pulses with product differentiation and availability of a wider range of products might be useful. Greater product differentiation is possible with technology upgrading and larger-scale processing, both of which seem to be lacking in the pulse processing sector, as highlighted in Chapter 5. Chapter 7 brings out these points in connection with convergent innovation. The Tata i-Shakti pulses, which are differentiated by the attribute of being unpolished, are an example where consumption has been claimed to have been positively affected. In short, processing is likely the next frontier in changing the consumption of pulses structurally.

Trade Policies

One of the important issues that this book highlights is related to India's pulses trade, specifically the structural break that has occurred over the previous decade. Pulses have now become India's second largest agricultural import, after edible oils. The rapid increase in pulse imports raises some fundamental questions. Should India strive hard for self-sufficiency in pulses, or should it let the other countries produce and export pulses to India? To put it another way: does India have a comparative advantage in pulses? Is there a level playing field between domestic producers and the exporters, or is there a suppressed comparative advantage in pulses because of domestic policies, infrastructure, and institutions over and above factor endowments?

Chapter 6 shows that it is not coincidental that disadvantaged crops, specifically oilseeds and pulses, show similar dynamics in trade. The shared pattern—of significant and persistent expansions of trade despite significant resources being committed to boost supply—indicates that there is a need to assess India's comparative advantage at least in some pulses. Recall that India, in general, has a highly protectionist trade policy, particularly in agricultural products. Pulses have been a clear exception to this policy, with tariffs being kept quite low (in several years at zero and in other years less than 10 percent). Hence, since India likely does not have a comparative advantage in pulses, a greater reliance on trade might be suggested. However, the question still remains whether reliance on external suppliers for pulses is advisable or even possible, given the size of India's import demand and the limited set of producers in the world.

The case of the pulses trade is unique in many respects. Even though trade has been expanding, most of the activity is concentrated on the intensive margin, as Chapter 6 shows. Only a limited number of countries have consistently been exporting pulses to India, and this concentration among a few exporters becomes even more pronounced in the case of specific pulses. It is only recently that the set of exporters has been expanding, with African countries joining in as exporters. For example, the latest entrant to the exporter pool is Sudan, which started exporting pigeon pea to India in 2014. This characteristic of a small set of exporters has clear implications for long-term trade policy in pulses. In June 2016 the government of India announced a policy to explore leasing land in African countries and Myanmar in order to grow pulses for export back to India, a unique move that has not been practiced by India for any other crop.

Overall, since there are only a few exporters and agricultural exports are subject to weather shocks (more so with climate change), depending heavily on trade to meet demand needs can sometimes lead to difficulties. Chapter 6 shows that shocks in some countries (like Canada for yellow pea) can lead to the emergence of some new exporters (for example, the United States), but such cases are few. Yet there is no indication from the analyses in this book that India has reached a level of import penetration in pulses without diversity in the set of exporters. If anything, it seems that pulse imports are playing a role below their potential in cooling domestic prices. Chapter 6 shows that while trade has been playing a prolonged role in cooling, it has been limited in its effect on bringing down prices. Though not analyzed in this book, the literature does show that the limited impact of liberalized trade in India at the border is often due to domestic factors (Kumar, Roy, and Gulati 2010). Factors like the quality of infrastructure and procedural delays play an important role in transmission, though to what extent they play a role in pulses remains an important area for further research.

In the case of pulses, as domestic production and consumption areas have separated in many instances, a major lesson that emerges from the studies is the importance of spatial integration of domestic markets. The *spatial integration* refers to the degree of price uniformity observed between different regions within a country and depends on the ease with which goods can be moved from a surplus region to a deficit region. Spatially integrated markets ensure that local shortages do not translate into a sharp rise in prices forcing a reduction in consumption. In the case of pulses, this has become quite important as there have been significant regional shifts in production over time, as Chapter 3 illustrates. Since production is dispersed and consumption (by variety) is concentrated, spatial integration of markets is becoming increasingly important for pulses in India. The benefits of trade liberalization in pulses would be maximized only when domestic markets are integrated (see Kumar, Roy, and Gulati 2010 for this argument).

Domestic Market Structures

One of the persistent complaints about private traders in the agricultural markets is that they have relatively more market power than either farmers or consumers. Because of this market power, in their transactions with farmers, traders have oligopsony powers; in their transactions with consumers, they have oligopoly powers. This market power can be limited if the buyer is not a direct consumer but a larger entity—for example, a processor. When traders sell to retailers, this allows them to be price-fixers and not price-takers in both transactions. The traders' margin is often high in commodities like pulses, and farmers are often subjected to receiving prices well below the market price (as outlined in Chapters 3 and 5). Mechanisms that promote direct marketing and create farmer collectives to improve their bargaining power can help in improving the price and income realization of the farmers. The government has often tried to break the hold of these traders in the market for pulses, because they indulge in hoarding and black marketing and do not allow farmers to realize the benefit when prices are high.

The Way Forward

As the discussion in this book brings out, government's primary role in the pulse sector should be in investing in infrastructure development and technology research and dissemination that would promote pulses' production and deepen markets. In this context, the pricing policy of pulses relative to other crops warrants further research and assessment. Also, government's role in pulses could be more proactive in facilitating private-sector participation—for example, in technology development. Private corporations may be encouraged to become more involved in the pulse sector. Some of the models discussed in Chapters 5 and 7, which are in their nascent stages, may offer some promise, although final judgment on the viability and scalability must await rigorous assessment. Most of these models seem to involve some form of public-private partnership and often involve the engagement of pulse processors. In some cases, the arrangement also involves providing more differentiated products.

Increasing Nutritional Intake from Pulses

Finally, steps need to be taken to help pulses meet their full nutrition potential. The decreasing preference for pulses, particularly among youths and children, is leading to a reduction in dietary protein intake from pulses. To increase protein intake, some novel initiatives are being taken by developing new pulse-based products. One such initiative is the Pulse Innovation Partnership (PIP), a global alliance of public and private organizations, civil society agencies, and academia, which aspires to partner with small, medium-size, and large food companies in multiple geographies, providing them with need-based knowledge services for innovation and marketing, opening up a large innovation channel for pulse-based products. The other nutrition-focused project is the Odisha Pilot as discussed in Chapter 7. This new model of development aims to make nutrition and healthcare priorities along with economic growth in the rural areas of Odisha. How far these projects deliver will have a bearing on the types of models to adopt for a comprehensive system for pulses for nutrition and health and environmental sustainability.

References

Asare-Marfo, D., E. Birol, B. Karandikar, D. Roy, and S. Singh. 2010. "Varietal Adoption of Pearl Millet (bajra) in Maharashtra and Rajasthan, India: A Summary Paper." HarvestPlus project report. Washington, DC: International Food Policy Research Institute (IFPRI).

Birthal, P. S., S. Kumar, D. S. Negi, and D. Roy. 2015. "The Impacts of Information on Returns from Farming: Evidence from a Nationally Representative Farm Survey in India." *Agricultural Economics* 46 (4): 549–561.

Chakraborty, S., A. Kishore, and D. Roy. 2016. "Effectiveness of Food Subsidies in Raising Healthy Food Consumption: Public Distribution of Pulses in India." IFPRI Discussion Paper 1523. Washington, DC: International Food Policy Research Institute (IFPRI).

Feder, G., R. Birner, and J. R. Anderson. 2011. "The Private Sectors Role in Agricultural Extension Systems: Potential and Limitations." *Journal of Agribusiness in Developing and Emerging Economies* 1: 31–54.

India, Ministry of Agriculture. 2009. "Report of Expert Group on Pulses." New Delhi: Department of Agriculture and Cooperation.

Joshi, P. K., and P. Kumar. 2011. "Food Demand and Supply Projections for India: 2010–2030." In *Trade, Agricultural Policies and Structural Changes in India's Agrifood System: Implications for National and Global Markets*. Report for TAPSIM Project KBBE212617. New Delhi: International Food Policy Research Institute.

Kumar, A. G., D. Roy, and A. Gulati. 2010. *Liberalizing Food Grains Markets: Experiences, Impacts, and Lessons from South Asia*. New Delhi: Oxford University Press.

NAAS (National Academy of Agricultural Sciences). 2013. "Improving Productivity of Rice Fallows." Policy Paper 64. New Delhi: NAAS.

Niranjan, S. K. D. F. F., M. C. S. Bantilan, P. K. Joshi, and K. B. Saxena. 1998. "Evaluating Hybrid Technology for Pigeon Pea." In *Proceedings of the International Workshop on Joint Impact Assessment of NARS/ICRISAT Technologies for the Semi-Arid Tropics*, 231–240. ICRISAT Center, Patancheru, India, December 2–4, 1996.

Pande, S., M. Sharma, and N. Ghosh. 2012. "Role of Pulses in Sustaining Agricultural Productivity in the Rainfed Rice-Fallow Lands of India in Changing Climatic Scenario." Patancheru, India: International Crops Research Institute for the Semi-Arid Tropics (ICRISAT).

Singh, N. P., and M. C. Saxena. 2016. "Towards Self-Sufficiency of Pulses in India." Base paper for National Academy of Agricultural Sciences (NAAS) brainstorming workshop on April 8, 2016. New Delhi.

Vyas, V. S. 2003. "Market Reforms in Indian Agriculture." Paper presented at Jawahar Nehru University NU-IFPRI Workshop on Economic and Agricultural Reforms in China and India, New Delhi, March 25–26.

Contributors

Raj Chandra (rchandra@iastate.edu) was a research analyst at the International Food Policy Research Institute, New Delhi. Currently, Chandra is a graduate student at Iowa State University.

Laurette Dubé (laurette.dube@mcgill.ca) is a full professor and holds the James McGill Chair of Consumer and Lifestyle Psychology and Marketing at the Desautels Faculty of Management of McGill University.

Srivardhini K. Jha (vardhini@gmail.com) is a research scientist at CGIAR/McGill Centre for the Convergence of Health and Economics.

P. K. Joshi (p.joshi@cgiar.org) is the director of the South Asia Office of the International Food Policy Research Institute, New Delhi.

Praduman Kumar (pkumariari@gmail.com) is the former head, Division of Agricultural Economics, Indian Agriculture Research Institute, New Delhi.

John McDermott (j.mcdermott@cgiar.org) is director of the CGIAR Research Program on Agriculture for Nutrition and Health of the International Food Policy Research Institute.

B. Mishra (b9mishra@gmail.com) is the former vice chancellor at S.K. University of Agricultural Sciences and Technology of Jammu, J & K, and former director, Directorate of Wheat Research, Karnal, and Directorate of Rice Research, Hyderabad.

Akanksha Negi (a.negi@cgiar.org) was a senior research assistant at the International Food Policy Research Institute, New Delhi. Currently, she is a graduate student at Michigan State University.

Shinoj Parappurathu (shinoj.p@icar.org.in) is an Agricultural Research Service (ARS) Scientist at the Indian Council of Agricultural Research, Central Marine Fisheries Research Institute (ICAR-CMFRI) in Kozhikode, Kerala, India.

Devesh Roy (d.roy@cgiar.org) is a research fellow in the Markets, Trade, and Institutions Division of the International Food Policy Research Institute, New Delhi.

Inba Sekar (economist.inba@gmail.com) is a scientist at the Indian Agriculture Research Institute, New Delhi.

Index

Note: Page numbers followed by "f" or "t" refer to figures or tables respectively.

Made in United States
North Haven, CT
04 November 2026

21719200R00157